Lecture Notes in Computer Science 14936

Founding Editors

Gerhard Goos
Juris Hartmanis

Editorial Board Members

Elisa Bertino, *Purdue University, West Lafayette, IN, USA*
Wen Gao, *Peking University, Beijing, China*
Bernhard Steffen ⓘ, *TU Dortmund University, Dortmund, Germany*
Moti Yung ⓘ, *Columbia University, New York, NY, USA*

The series Lecture Notes in Computer Science (LNCS), including its subseries Lecture Notes in Artificial Intelligence (LNAI) and Lecture Notes in Bioinformatics (LNBI), has established itself as a medium for the publication of new developments in computer science and information technology research, teaching, and education.

LNCS enjoys close cooperation with the computer science R & D community, the series counts many renowned academics among its volume editors and paper authors, and collaborates with prestigious societies. Its mission is to serve this international community by providing an invaluable service, mainly focused on the publication of conference and workshop proceedings and postproceedings. LNCS commenced publication in 1973.

Harold Mouchère · Anna Zhu
Editors

Document Analysis and Recognition – ICDAR 2024 Workshops

Athens, Greece, August 30–31, 2024
Proceedings, Part II

Editors
Harold Mouchère
Nantes Université
Nantes, France

Anna Zhu
Wuhan University of Technology
Wuhan, China

ISSN 0302-9743   ISSN 1611-3349 (electronic)
Lecture Notes in Computer Science
ISBN 978-3-031-70641-7   ISBN 978-3-031-70642-4 (eBook)
https://doi.org/10.1007/978-3-031-70642-4

© The Editor(s) (if applicable) and The Author(s), under exclusive license to Springer Nature Switzerland AG 2024

This work is subject to copyright. All rights are solely and exclusively licensed by the Publisher, whether the whole or part of the material is concerned, specifically the rights of translation, reprinting, reuse of illustrations, recitation, broadcasting, reproduction on microfilms or in any other physical way, and transmission or information storage and retrieval, electronic adaptation, computer software, or by similar or dissimilar methodology now known or hereafter developed.
The use of general descriptive names, registered names, trademarks, service marks, etc. in this publication does not imply, even in the absence of a specific statement, that such names are exempt from the relevant protective laws and regulations and therefore free for general use.
The publisher, the authors and the editors are safe to assume that the advice and information in this book are believed to be true and accurate at the date of publication. Neither the publisher nor the authors or the editors give a warranty, expressed or implied, with respect to the material contained herein or for any errors or omissions that may have been made. The publisher remains neutral with regard to jurisdictional claims in published maps and institutional affiliations.

This Springer imprint is published by the registered company Springer Nature Switzerland AG
The registered company address is: Gewerbestrasse 11, 6330 Cham, Switzerland

If disposing of this product, please recycle the paper.

Foreword

We are honoured to welcome you to the proceedings of ICDAR 2024, the 18th IAPR International Conference on Document Analysis and Recognition, which took place in Athens, the beautiful and historic capital of Greece. ICDAR 2024 marked the start of the annual basis for the ICDAR series.

ICDAR 2024 was the 18th edition of a longstanding conference series that has come of age, sponsored by the International Association for Pattern Recognition (IAPR). It is the premier international event for scientists and practitioners in document analysis and recognition. This field continues to play an important role in document understanding and recognition.

The IAPR TC10/11 technical committees endorse the conference. The very first ICDAR was held in St. Malo, France in 1991, followed by Tsukuba, Japan (1993), Montreal, Canada (1995), Ulm, Germany (1997), Bangalore, India (1999), Seattle, USA (2001), Edinburgh, UK (2003), Seoul, South Korea (2005), Curitiba, Brazil (2007), Barcelona, Spain (2009), Beijing, China (2011), Washington, DC, USA (2013), Nancy, France (2015), Kyoto, Japan (2017), Sydney, Australia (2019), Lausanne, Switzerland (2021) and San Jose, USA (2023).

Keeping with its tradition from past years, ICDAR 2024 featured a three-day main conference, including several competitions to challenge the field and a pre-conference slate of workshops, tutorials, and a doctoral consortium. The conference was held in Athens, Greece on September 2–4, 2024, and the pre-conference tracks on August 30 till September 1, 2024.

The highlights of the conference included keynote talks by the recipient of the IAPR/ICDAR Outstanding Achievements Award, and distinguished speakers: Jürgen Schmidhuber, Director of the AI Initiative at KAUST, Swiss AI Lab IDSIA, and the University of Lugano, Switzerland; Maria Kamilaki, Acting Director-General of e-Administration, Library & Publications of the Hellenic Parliament, Greece; and Cheng-Lin Liu, State Key Laboratory of Multimodal Artificial Intelligence Systems, Institute of Automation of the Chinese Academy of Sciences, China.

A total of 263 papers were submitted to the main conference (plus 35 papers to the ICDAR-IJDAR journal track), with 52 papers accepted for oral presentation (plus 17 IJDAR track papers) and 92 for poster presentation. We would like to express our deepest gratitude to our Program Committee Chairs, featuring three distinguished researchers from academia, Elisa Barney Smith, Liangrui Peng, and Marcus Liwicki, who did a phenomenal job in overseeing a comprehensive reviewing process and who worked tirelessly to put together a very thoughtful and interesting technical program for the main conference. We are also very grateful to the members of the Program Committee for their high-quality peer reviews. We extend our gratitude to our competition chairs, George Retsinas and Xiang Bai, for overseeing the competitions.

The pre-conference featured 6 excellent workshops, 4 value-filled tutorials, and the doctoral consortium. We would like to thank Harold Mouchère and Anna Zhu, the workshop chairs, Vincent Christlein and Alicia Fornés, the tutorial chairs, and KC Santosh and Andreas Fischer, the doctoral consortium chairs, for their efforts in putting together a wonderful pre-conference program. We would like to thank and acknowledge the hard work put in by our Publication Chairs, Giorgos Sfikas and Christophoros Nikou, who worked diligently to compile the camera-ready versions of all the papers and organize the conference proceedings with Springer. Many thanks are also due to our sponsorship, awards, industry, and publicity chairs for their support of the conference.

Finally, we would like to thank our many financial sponsors for their support and the conference attendees and authors, for helping make this conference a success. We sincerely hope that all attendees had an enjoyable conference, a wonderful stay in Athens, and fruitful academic exchanges with their colleagues.

August 2024

Basilis Gatos
Vassilis Katsouros
Foteini Simistira Liwicki

Preface

Welcome to the proceedings of the 18th International Conference on Document Analysis and Recognition (ICDAR 2024). ICDAR is the premier international event for scientists and practitioners involved in document analysis and recognition. This iteration is the first iteration in an even year, marking the beginning of the series becoming an annual event.

This year, we received 263 conference paper submissions. In order to create a high-quality scientific program for the conference, we recruited 159 regular and 32 senior program committee (PC) members. Each paper received at least 2 single-blind reviews, with most papers receiving 3 or more reviews. In addition, senior PC members who oversaw the review phase for typically 8–10 submissions took care of consolidating reviews and suggested paper decisions in their meta-reviews. Based on the information provided in both the reviews and the prepared meta-reviews we PC Chairs then selected 144 submissions (54.8%) for inclusion in the scientific program of ICDAR 2024. From the accepted papers, 52 were selected for oral presentation, and 92 for poster presentation.

In addition to the papers submitted directly to ICDAR 2024, we continued the tradition of teaming up with the International Journal of Document Analysis and Recognition (IJDAR) and organized a special journal issue. The journal-track submissions underwent the same rigorous review process as regular IJDAR submissions. The ICDAR PC Chairs served as Guest Editors and oversaw the review process. From the 35 manuscripts submitted to the journal track, 17 were accepted and were published in a Special Issue of IJDAR entitled "Advanced Topics of Document Analysis and Recognition." In addition, all papers accepted in the journal track were included as oral presentations in the conference program with a slightly extended presentation time (25 instead of 20 minutes including Q&A).

With the transition of ICDAR to an annual conference, the alternate year workshops DAS and ICFHR also made some changes. DAS decided to now be colocated with ICDAR. Handwritten text recognition is now a prominent topic in the DAR community, so the organizers and attendees of ICFHR 2023 decided to become part of the main ICDAR conference. Two sessions in ICDAR 2024 were on Frontiers of Handwriting Recognition and a third was devoted to Chinese text recognition. The popular Workshop on Historical Document Imaging and Processing (HIP) decided to remain an alternate year workshop, but in odd years. The main ICDAR conference had 2 sessions devoted to Historical Document Analysis. Scene text recognition, music, tables and charts continue to be topics of interest. Visual Question Answering, Document Understanding and NLP are rising topics gaining enough papers to devote sessions to them. As over the years many new machine learning techniques have been developed by researchers in the DAR community, a track also was devoted to papers that focused more on the methods than on any particular application. This year, nine scientific competitions were held in conjunction with ICDAR. A session was devoted to presenting the results.

As ICDAR 2024 was held with in-person attendance, all papers were presented by their authors during the conference. Exceptions were only made for authors who could not attend the conference for unavoidable reasons. Such oral presentations were then provided by synchronous video presentations. Posters of authors that could not attend were presented by recorded teaser videos, in addition to the physical posters.

Three keynote talks were given by Jürgen Schmidhuber, Director of the AI Initiative at KAUST, Swiss AI Lab IDSIA, and University of Lugano, Switzerland; Maria Kamilaki, Acting Director-General of e-Administration, Library & Publications of the Hellenic Parliament, Greece; and Cheng-Lin Liu, State Key Laboratory of Multimodal Artificial Intelligence Systems, Institute of Automation of the Chinese Academy of Sciences, China and the recipient of the IAPR/ICDAR Outstanding Achievements Award. We thank them for the valuable insights and inspiration that their talks provided for participants.

Finally, we would like to thank everyone who contributed to the preparation of the scientific program of ICDAR 2024, namely the authors of the scientific papers submitted to the journal track and directly to the conference, reviewers for journal-track papers, and both our regular and senior PC members. We also thank the Springer staff and the ICDAR 2024 Publication Chairs, who oversaw the creation of these proceedings.

August 2024

Elisa Barney Smith
Marcus Liwicki
Liangrui Peng

Foreword from ICDAR 2024 Workshop Chairs

We extend a warm welcome to the proceedings of the ICDAR 2024 Workshops, which were organized as part of the 18th International Conference on Document Analysis and Recognition (ICDAR) held in Athens, Greece from August 30 to September 4, 2024. The workshops were conducted prior to the commencement of the main conference, on the 30th and 31st of August, 2024. All of the workshops were conducted in person, as was the main conference.

The ICDAR conference comprised five workshops, which addressed a range of document image analysis and recognition topics, as well as related subjects such as natural language processing, computational paleography, and digital humanities. In total, 46 papers were submitted for consideration, and 30 were accepted, representing a global acceptance rate of 65%.

This volume compiles the edited papers from the five workshops. We extend our sincerest gratitude to the ICDAR general chairs for entrusting us with the responsibility of organizing the workshops and to the publication chairs for their invaluable assistance in publishing this volume. We also express our profound appreciation to the workshop organizers for their invaluable contributions to this pivotal event in our field. Finally, we acknowledge and thank all the workshop presenters and authors.

September 2024

Harold Mouchère
Anne Zhu

Organization

Organizing Committee

General Chairs

Basilis Gatos — NCSR "Demokritos", Greece
Vassilis Katsouros — Athena Research Center, Greece
Foteini Simistira Liwicki — Luleå University of Technology, Sweden

Program Committee Chairs

Elisa Barney Smith — Luleå University of Technology, Sweden
Marcus Liwicki — Luleå University of Technology, Sweden
Liangrui Peng — Tsinghua University, China

Workshop Chairs

Harold Mouchère — Nantes Université, France
Anna Zhu — Wuhan University of Technology, China

Competition Chairs

George Retsinas — National Technical University of Athens, Greece
Xiang Bai — Huazhong Univ. of Sci. & Technology, China

Tutorial Chairs

Vincent Christlein — University of Erlangen-Nuremberg, Germany
Alicia Fornés — Universitat Autònoma de Barcelona, Spain

Publication Chairs

Giorgos Sfikas — University of West Attica, Greece
Christophoros Nikou — University of Ioannina, Greece

Doctoral Consortium Chairs

Andreas Fischer	Univ. of App. Sci. & Arts Western Switzerland, Switzerland
K. C. Santosh	University of South Dakota, USA

Awards Chairs

Michael Blumenstein	University of Technology Sydney, Australia
Ioannis Pratikakis	Democritus University of Thrace, Greece

Posters/Demo Chairs

Umapada Pal	Indian Statistical Institute, India
Momina Moetesum	National University of Sciences & Technology, Pakistan
Kenny Davila	DePaul University, USA

Sponsorship Chairs

Markus Weber	Wacom, USA
Xu-Cheng Yin	University of Science and Technology Beijing, China

Industry Chairs

Dimosthenis Karatzas	Universitat Autònoma de Barcelona, Spain
Srirangaraj Setlur	University at Buffalo, USA
Errui Ding	Baidu Inc., China

Local Organization Chairs

Anastasios Kesidis	University of West Attica, Greece
Kosmas Kritsis	Athena Research Center, Greece
Elena Galifianaki	NCSR "Demokritos", Greece
Pelagia Drosaki	NCSR "Demokritos", Greece

Publicity Chairs

Panagiotis Kaddas	NCSR "Demokritos", Greece
Vassilis Papavassiliou	Athena Research Center, Greece
Elena Galifianaki	NCSR "Demokritos", Greece

Program Committee

Senior Program Committee Members

Apostolos Antonacopoulos	University of Salford, UK
Anurag Bhardwaj	eBay Research Labs, USA
Michael Blumenstein	University of Technology, Sydney, Australia
Jean-Christophe Burie	L3I - Université de La Rochelle, France
Bertrand Coüasnon	Irisa/Insa, France
Mickaël Coustaty	Laboratoire L3i - La Rochelle Université, France
David Doermann	University at Buffalo, USA
Véronique Eglin	LIRIS-INSA de Lyon, France
Gernot Fink	TU Dortmund, Germany
Andreas Fischer	University of Fribourg, Switzerland
Alicia Fornés	Computer Vision Center, UAB, Spain
Liangcai Gao	Peking University, China
Nicholas Howe	Smith College, USA
C. V. Jawahar	CVIT, IIIT, Hyderabad, India
Lianwen Jin	South China University of Technology, China
Dimosthenis Karatzas	CVC, Universitat Autónoma de Barcelona, Spain
Koichi Kise	Osaka Metropolitan University, Japan
Bart Lamiroy	Université de Reims Champagne-Ardenne, France
Cheng-Lin Liu	CASIA, China
Lu Liu	Lazada, Singapore
Josep Llados	CVC, Universitat Autónoma de Barcelona, Spain
Daniel Lopresti	Lehigh University, USA
R. Manmatha	University of Massachusetts, Amherst, USA
Angelo Marcelli	Università di Salerno, Italy
Simone Marinai	University of Florence, Italy
Jean-Marc Ogier	University of La Rochelle, France
Wataru Ohyama	Tokyo Denki University, Japan
Marçal Rusiñol	AllRead Machine Learning Technologies, Spain
Robert Sablatnig	TU Wien, Austria
Faisal Shafait	National University of Sciences and Technology, Pakistan
Seiichi Uchida	Kyushu University, Japan
Jerod Weinman	Grinnell College, USA
Richard Zanibbi	Rochester Institute of Technology, USA
Yu Zhou	Nankai University, Japan

Program Committee Members

Irfan Ahmad
Alireza Alaei
Musab Al-Ghadi
Eric Anquetil
Vlad Atanasiu
Muhammad Naseer Bajwa
Byron Bezerra
Ujjwal Bhattacharya
Jean-Luc Bloechle
Alceu Britto
Rina Buoy
Jorge Calvo-Zaragoza
Cristina Carmona-Duarte
Sukalpa Chanda
Clément Chatelain
Bidyut B. Chaudhuri
Joseph Chazalon
Shanxiong Chen
Jin Chen
Youssouf Chherawala
Vincent Christlein
Christian Clausner
Mark Clement
Florence Cloppet
Kenny Davila
Claudio De Stefano
Abhisek Dey
Sounak Dey
Antoine Doucet
Fadoua Drira
Mounîm A. El Yacoubi
Jonathan Fabrizio
Francesco Fontanella
Yasuhisa Fujii
Akio Fujiyoshi
Rajib Ghosh
Romain Giot
Lluis Gomez
Petra Gomez-Krämer
Daichi Haraguchi
Sheng He
Nina S. T. Hirata
Qiang Huo
Donato Impedovo
Brian Kenji Iwana
Maham Jahangir
Aashi Jain
Mohammed Javed
Jobin K. V.
Ehsanollah Kabir
Karim Kalti
Lei Kang
Slim Kanoun
Christopher Kermorvant
Yousri Kessentini
Florian Kleber
Pramod Kompalli
Aurélie Lemaitre
Hongjun Li
Zhouhui Lian
Lingyu Liang
Minghui Liao
Laurence Likforman
Rafael Lins
Chang Liu
Yuliang Liu
Muhammad Muzzamil Luqman
Nam Tuan Ly
Sriganesh Madhvanath
Nishatul Majid
Carlos David Martinez Hinarejos
Maroua Mehri
Carlos Mello
Ronaldo Messina
Evangelos Milios
Zuheng Ming
Tomo Miyazaki
Momina Moetesum
Hussein Mohammed
Ajoy Mondal
Harold Mouchère
Shobharani N.
Nibal Nayef
Clemens Neudecker
Hung Tuan Nguyen
Shinichiro Omachi

Umapada Pal
Shivakumara Palaiahnakote
Thierry Paquet
Mohammad Tanvir Parvez
Antonio Parziale
Marco Peer
Dezhi Peng
Vincent Poulain D'Andecy
Ioannis Pratikakis
Irina Rabaev
Jean-Yves Ramel
Oriol Ramos-Terrades
Frédéric Rayar
Kaspar Riesen
Christophe Rigaud
Verónica Romero
Henry A. Rowley
Joan Andreu Sanchez
Ravi Kiran Sarvadevabhatla
Martin Schall
Amina Serir
Anuj Sharma
Ying Sheng
Nicolas Sidère
Steven Simske
Sukhdeep Singh
Daniel Stoekl Ben Ezra
Tonghua Su
Xiangdong Su
Suresh Sundaram
Salvatore Tabbone
Sandeep Tata
Christopher Tensmeyer
Kengo Terasawa
Iuliia Tkachenko
Ruben Tolosana
Alejandro Toselli
Xiao Tu
Oliver Tüselmann
Huy Quang Ung
Szilard Vajda
Ernest Valveny
Ekta Vats
Ruben Vera-Rodriguez
Enrique Vidal

Lars Vögtlin
Yanwei Wang
Qiu-Feng Wang
Da-Han Wang
Yang Xue
Chun Yang
Mingkun Yang
Berrin Yanikoglu
Fei Yin
Qi Zeng
Heng Zhang
Yanming Zhang
Guangwei Zhang
Yuchen Zheng
Anna Zhu
Majid Ziaratban
Chandranath Adak
Oluwatosin Adewumi
Akshay Agarwal
Peeta Basa Pati
Khadiravana Belagavi
Asma Bensalah
Mohammad Idrees Bhat
Mélodie Boillet
Victoria Bourgeais
Iheb Brini
Francisco J. Castellanos
Francesco Castro
Apurba Chakraborty
Xu Chen
Denis Coquenet
Simon Corbillé
Aravinda Cv
Tiziana D'Alessandro
Avijit Dasgupta
Julien Delaunay
Vincenzo Dentamaro
Shubhang Desai
Alessandra Scotto di Freca
Moises Diaz
Ray Ding
Kalvin Dobler
Biyi Fang
Yuhang Fu
Gilad Fuchs

Cristiano Garcia
Vincenzo Gattulli
Loann Giovannangeli
Nathalie Girard
Tongkun Guan
Ahmed Hamdi
Raphaela Heil
Andre Hochuli
Kai Hu
Ludvig Hult
Syed Mohammad Baqir Husain
Nushrat Hussain
Aman Jaiswal
Mahdi Jampour
Nanfeng Jiang
Wang Jiawei
Michael Jungo
Wafa Khlif
Florian Kordon
Omar Krichen
Ahana Kundu
Songze Li
Zhixin Liu
Dongliang Luo
Puneet Mathur
Lin Meng
Elmokhtar Mohamed Moussa
Omar Moured
Emanuele Nardone
Emanuel Orler
Glen Pouliquen
Zhidong Qiao
Xingming Qu
Sachin Raja
Bulla Rajesh
Yann Ricquebourg
Antonio Ríos-Vila
Hugo Romat
Anna Scius-Bertrand
Gianfranco Semeraro
Mathias Seuret
Yilin Shi
Yongxin Shi
Mohamed Ali Souibgui
Yann Soullard
Maksym Taranukhin
Solène Tarride
Stacey Taylor
Vishvesh Trivedi
David Villanova-Aparisi
Manuel Villarreal Ruiz
Jiawei Wang
Xuewen Wang
Minghui Xia
Yejing Xie
Fuxiang Yang
Zhenhua Yang
Yan Zheng
Peijun Zou

Contents – Part II

IWCP

An Interpretable Deep Learning Approach for Morphological Script Type Analysis 3
   Malamatenia Vlachou-Efstathiou, Ioannis Siglidis, Dominique Stutzmann, and Mathieu Aubry

Detecting and Deciphering Damaged Medieval Armenian Inscriptions Using YOLO and Vision Transformers 22
   Chahan Vidal-Gorène and Aliénor Decours-Perez

Optimizing HTR and Reading Order Strategies for Chinese Imperial Editions with Few-Shot Learning 37
   Marie Bizais-Lillig, Chahan Vidal-Gorène, and Boris Dupin

Mind the Gap: Analyzing Lacunae with Transformer-Based Transcription 57
   Jaydeep Borkar and David A. Smith

NeuroPapyri: A Deep Attention Embedding Network for Handwritten Papyri Retrieval 71
   Giuseppe De Gregorio, Simon Perrin, Rodrigo C. G. Pena, Isabelle Marthot-Santaniello, and Harold Mouchère

MONSTERMASH: Multidirectional, Overlapping, Nested, Spiral Text Extraction for Recognition Models of Arabic-Script Handwriting 87
   Danlu Chen, Jacob Murel, Taimoor Shahid, Xiang Zhang, Jonathan Parkes Allen, Taylor Berg-Kirkpatrick, and David A. Smith

A New Framework for Error Analysis in Computational Paleographic Dating of Greek Papyri 102
   Giuseppe De Gregorio, Lavinia Ferretti, Rodrigo C. G. Pena, Isabelle Marthot-Santaniello, Maria Konstantinidou, and John Pavlopoulos

Automated Dating of Medieval Manuscripts with a New Dataset 119
   Boraq Madi, Nour Atamni, Vasily Tsitrinovich, Daria Vasyutinsky-Shapira, Jihad El-Sana, and Irina Rabaev

Image-to-Image Translation Approach for Page Layout Analysis
and Artificial Generation of Historical Manuscripts 140
    Chahan Vidal-Gorène and Jean-Baptiste Camps

VINALDO

A Multimodal Framework For Structuring Legal Documents 163
    Thibaud Real and Pauline Chavallard

Reformulating Key-Information Extraction as Next Sentence Prediction
for Hierarchical Data 175
    Ashish Kubade, Prathyusha Akundi, and Bilal Arif Syed Mohd

HPSegNet: A Method for Handwritten and Printed Text Separation
in Document Images 184
    Yu Chao, Changsong Liu, Liangrui Peng, and Yanwei Wang

Ablation Study of a Multimodal Gat Network on Perfect Synthetic
and Real-world Data to Investigate the Influence of Language Models
in Invoice Recognition 199
    Lukas-Walter Thiée

Author Index 213

Contents – Part I

ADAPDA

Domain Adaptation for Handwriting Trajectory Reconstruction from IMU Sensors 3
  Florent Imbert, Romain Tavenard, Yann Soullard, and Eric Anquetil

TrOCR Meets Language Models: An End-to-End Post-correction Approach ... 12
  Yung-Hsin Chen and Phillip B. Ströbel

LayeredDoc: Domain Adaptive Document Restoration with a Layer Separation Approach 27
  Maria Pilligua, Nil Biescas, Javier Vazquez-Corral, Josep Lladós, Ernest Valveny, and Sanket Biswas

Normalized vs Diplomatic Annotation: A Case Study of Automatic Information Extraction from Handwritten Uruguayan Birth Certificates 40
  Natalia Bottaioli, Solène Tarride, Jérémy Anger, Seginus Mowlavi, Marina Gardella, Antoine Tadros, Gabriele Facciolo, Rafael Grompone von Gioi, Christopher Kermorvant, Jean-Michel Morel, and Javier Preciozzi

ARPC

Diminutives in Political Discourse – The Case of Serbian and Slovenian 59
  Milena Oparnica

Loghi: An End-to-End Framework for Making Historical Documents Machine-Readable 73
  Rutger van Koert, Stefan Klut, Tim Koornstra, Martijn Maas, and Luke Peters

Open Parliamentary Data as a Tool for Linguistic Research: Exploring the 'Greek Language Question' in the *Journal of Parliamentary Debates* 89
  Maria Kamilaki

Digitization of Written Parliamentary Questions from the Historical
Archive (1974–1977) of the Hellenic Parliament 103
    Fotios Fitsilis, Basilis Gatos, Konstantinos Palaiologos,
    Panagiotis Kaddas, Charalambis Kyrkos, Maria-Eleni Georgoulea,
    Yiannis Armenakis, Christina Tasouli, George Mikros,
    Olivier Rozenberg, and Eleni Kiousi

MANPU

Retrieving and Analyzing Translations of American Newspaper Comics
with Visual Evidence 125
    Jacob Murel and David A. Smith

Investigating Neural Networks and Transformer Models for Enhanced
Comic Decoding 138
    Eleanna Kouletou, Vassilis Papavassiliou, and Vassilis Katsouros

Comics Datasets Framework: Mix of Comics Datasets for Detection
Benchmarking 154
    Emanuele Vivoli, Irene Campaioli, Mariateresa Nardoni,
    Niccolò Biondi, Marco Bertini, and Dimosthenis Karatzas

A Comprehensive Gold Standard and Benchmark for Comics Text
Detection and Recognition 168
    Gürkan Soykan, Deniz Yuret, and Tevfik Metin Sezgin

Toward Accessible Comics for Blind and Low Vision Readers 198
    Christophe Rigaud, Jean-Christophe Burie, and Samuel Petit

Quantitative Evaluation Based on CLIP for Methods Inhibiting Imitation
of Painting Styles 216
    Motoi Iwata, Keito Okamoto, and Koichi Kise

Spatially Augmented Speech Bubble to Character Association via Comic
Multi-task Learning 231
    Gürkan Soykan, Deniz Yuret, and Tevfik Metin Sezgin

ComicBERT: A Transformer Model and Pre-training Strategy
for Contextual Understanding in Comics 257
    Gürkan Soykan, Deniz Yuret, and Tevfik Metin Sezgin

Author Index 283

IWCP

IWCP 2024 Preface

Computational paleography is an emerging field investigating new computational approaches for analyzing ancient documents. Paleography, understood as the study of ancient writing systems (scripts and their components) as well as their material (characteristics of the physical inscribed objects), can benefit greatly from recent technological advances in computer vision and instrumental analytics. Computational paleography, being truly interdisciplinary, creates opportunities for experts from different research fields to meet, discuss, and exchange ideas. Collaborations between manuscript specialists in the Humanities rarely overcome the chronological and geographical boundaries of each discipline. However, when it comes to applying optical, chemical, or computational analysis, these boundaries are often no longer relevant. On the other hand, computer scientists are keen to confront their methodologies with actual research questions based on solid data. Natural scientists working either on the physical properties of the written artefacts or on the production of their digital "avatar" are the third link in this chain of knowledge. In many cases, only a collaboration between experts from different communities can yield significant results.

In this workshop, we aimed to bring together specialists of the different research fields analyzing handwritten scripts in ancient artefacts. It mainly targeted computer scientists, natural scientists, and humanists involved in the study of ancient scripts. By fostering discussion between communities, it facilitated future interdisciplinary collaborations that tackle actual research questions on ancient manuscripts.

The third edition was held in-person on August 31, 2024 in Athens, in conjunction with ICDAR 2024. It was organized by Isabelle Marthot-Santaniello (Institute for Ancient Civilizations, Universität Basel, Switzerland) and Hussein Adnan Mohammed (Centre for the Study of Manuscript Cultures, Universität Hamburg, Germany). The Program Committee counted 19 members and was selected to reflect the interdisciplinary nature of the field.

For this third edition, we welcomed two kinds of contributions: full papers and abstracts. We received a total of 21 submissions (14 papers and 7 abstracts). Each full paper was reviewed by two members of the program committee via EasyChair and nine were accepted. A blind review was used for the full paper submissions. The seven abstract submissions were evaluated by the organizers, and were published separately by the organizers in a dedicated website .

August 2024
                                                            Isabelle Marthot-Santaniello
                                                                      Hussein Mohammed

An Interpretable Deep Learning Approach for Morphological Script Type Analysis

Malamatenia Vlachou-Efstathiou[1,2(✉)], Ioannis Siglidis[2], Dominique Stutzmann[1], and Mathieu Aubry[2]

[1] Institut de Recherche et d'Histoire des Textes, Paris, Île-de-France, France
{malamatenia.vlachou,dominique.stutzmann}@irht.cnrs.fr
[2] LIGM, Ecole des Ponts, University of Gustave Eiffel, CNRS, Marne-la-Vallée, France
{ioannis.siglidis,mathieu.aubry}@enpc.fr

Abstract. Defining script types and establishing classification criteria for medieval handwriting is a central aspect of palaeographical analysis. However, existing typologies often encounter methodological challenges, such as descriptive limitations and subjective criteria. We propose an interpretable deep learning-based approach to morphological script type analysis, which enables systematic and objective analysis and contributes to bridging the gap between qualitative observations and quantitative measurements. More precisely, we adapt a deep instance segmentation method to learn comparable character prototypes, representative of letter morphology, and provide qualitative and quantitative tools for their comparison and analysis. We demonstrate our approach by applying it to the *Textualis Formata* script type and its two subtypes formalized by A. Derolez: Northern and Southern *Textualis*.

Keywords: Latin Palaeography · Computer Vision · Palaeographical Analysis · Character Prototypes · Textualis Formata

1 Introduction

The concept of *script type* is of central importance to palaeography, which studies handwritten documents in relation to their context of production such as date, origin, and scribal hands, to enrich historical discourse. Adapting M. Parkes, we define a *script type* as "the model which the scribe has in their mind's eye when they write" [43,56], that is, if we restrict ourselves to characters, the set of prototypical forms of each character towards which they are working when they write. To discretize the *continuum* of handwritten forms, and establish script types and their classification criteria, the palaeographical method compares handwriting samples, describes, and analyzes the variations of letter forms. Palaeographers often resort to the idea of script types as *ideal prototypes* [6], with the delineation of artificial alphabets, i.e., sets of abstracted letter forms. Several typologies have been proposed and refined over the years. Most recently A. Derolez proposed a taxonomy for Gothic book scripts based on letter morphology [16], where some letters with distinctive visual elements serve as the basis for classification.

In this paper, we introduce a methodology that leverages deep learning for the analysis of morphological script types. More precisely, we learn aligned character prototypes from documents and present different methods for qualitative and quantitative analysis. This enables us to confront different documents to existing typologies, potentially adding nuance or complementing them. Indeed, existing typologies present persistent methodological issues [16,53,58], mainly the ambiguity arising from relying on a "global impression" to discern scripts, inconsistencies in nomenclature across scholarly traditions, and difficulties describing minute morphological differences using natural language [19,56]. These challenges underscore the potential benefits of methods such as ours, which could enhance palaeographical analysis through a systematic and objective approach, and facilitate the integration of quantitative measures with qualitative observations.

This is in line with the position of pioneers like Léon Gilissen [20,21] and others [15,37,39,42,46,52,59,66], who experimented with statistical measurements and the modeling of *measurable elements* of script. Such measurements of scripts pose significant challenges, such as defining a set of descriptors or discriminative handwriting features and ensuring comparable objects and magnitudes [55]. Contrary to classification tasks such as writer and geographical attribution, which are formulated as discriminative learning problems, script type analysis cannot be reduced to a classification problem [26,60,61] and adequate modeling of variations is crucial [57]. Indeed simply matching external samples to pre-defined script types does not help better understanding and questioning the classification criteria.

We thus propose a method for evidence-based paleography focusing on interpretability rather than script classification. Our key idea is to remain close to classical morphological approaches for defining taxonomies and introduce tools to model and analyze letter shapes automatically. We build on the Learnable Typewriter approach [51] and adapt it so that it can learn comparable character prototypes, which requires designing appropriate finetuning strategy and filtering. We then introduce visualizations and graphical tools, as well as an interpretable variability measure. To demonstrate how such tools can be leveraged for palaeographic analysis, we select a corpus in *Textualis Formata* and present a case study on the morphological analysis of its two subtypes, Northern and Southern *Textualis*.

Contribution. In summary, our main contributions are:

- the adaptation of a deep instance segmentation method for palaeographical script type analysis
- a methodology for homologous comparison of characters, including visualization, graphical, and quantitative tools
- a case study demonstrating how these tools can complement the classic taxonomy of A. Derolez [16] for the analysis of Northern and Southern *Textualis*.

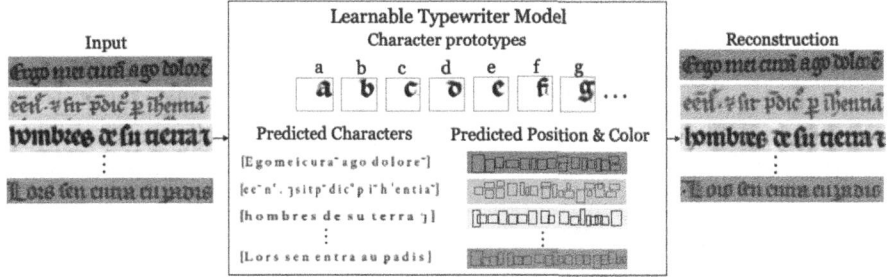

Fig. 1. The Learnable Typewriter Model learns to reconstruct text lines using a set of learned character prototypes. We demonstrate how the character prototypes can be used for palaeographic analysis.

2 Related Work

We first give an overview of works that develop quantitative methods for palaeographic analysis. We then present "prototype-based" approaches to document analysis, which, although not specifically developed for paleographic analysis, are the basis of our approach.

Quantitative Methods in Palaeography. In the past two decades, many automatic methods have been developed for writer or script classification [40, 54], using texture-based features [17,25,27,34,38,49,67], grapheme-based features [18,29,48,50] and deep learning classification approaches [7,9–11,28,64,65]. Some papers, such as [31], make a particular effort to build interpretable features or to visualize deep features responsible for the classification, but their interpretation remains limited.

Another branch of studies aims at producing interpretable outputs for palaeographic analysis of letter structure. The Information System for Graphological Identification [36], extracts the average shape of specific characters via curve and contour detection, standardizing orientation and size, for automatic hand comparison and writer identification. The Graphem project [39] focused specifically on script type features. [14] explores visually interpretable stroke analysis, by extracting connected components to create a strokes code book, and then grouping the strokes through graph coloring for categorization of elementary stroke shapes. Closest to us, focusing on entire letter form variations for script type analysis is the System of Palaeographical Inspection [1,8]. It generates an average character prototype by computing the centroid of semi-automatically segmented occurrences. The prototypes are used both for hierarchical clustering of similar hands and classification of external samples.

However, the results of these approaches can hardly be compared with minute traditional palaeographic analysis and do not provide complete automation. Instead, our idea is to build on methods that directly learn prototypical characters from documents and use them for actual paleographic analysis.

Prototype-based Approaches in Document Analysis. Early methods for document analysis [3,4,30,32,68] use variants of character template matching for analyzing documents. While their main goal is often to perform optical character recognition (OCR), such methods typically also produce finer outputs, such as character segmentation, and learn a character template or prototype. Similar approaches have thus been used for typographical analysis of early prints [24,33,47]. This type of approach has recently been revisited with deep learning tools by the Learnable Typewriter approach [51]. We build on this method and describe it in the next section.

3 Approach

3.1 Learning Comparable Character Prototypes

The Learnable Typewriter. We build our approach on the Learnable Typewriter model [51], visualized in Fig. 1. This deep learning model learns to reconstruct text lines by compositing a set of *character prototypes* on a simple background. Given as input the image of a line, it predicts the color of the background, the characters used in the line, and for each character, its position and color. The character prototypes are also learned by the model, and each instance of a character is reconstructed with the exact same prototype. The model can be trained, as in our experiments, using a set of text line images with their transcriptions.

Each character prototype is a grayscale image and can be thought of as the average shape of all occurrences of a character in the training data, standardized for color, size, and position. Therefore, training a Learnable Typewriter model on a particular corpus, such as one corresponding to a specific script or handwriting style, will yield the average shape of each character without the need for manual selection of specific character samples, annotation of character positions, or binarization.

Finetuning Character Prototypes. We propose to compare different documents and different scripts by comparing the character prototypes learned on various corpora. However, directly comparing prototypes learned by different models is not possible, because they are not aligned. Our solution is first to learn a *reference model* using a reference corpus - in our case study, a set of documents in *Textualis Formata* - then finetune the model to reconstruct selected documents, keeping all network parameters frozen except those that only impact the prototypes. Since the positioning, scaling, and coloring of the prototypes are shared, the prototypes will remain aligned and can be directly compared, such as by computing their difference. We define a single reference corpus to obtain reference prototypes, and then finetune them on multiple specific corpora, that may or may not be part of the reference corpus.

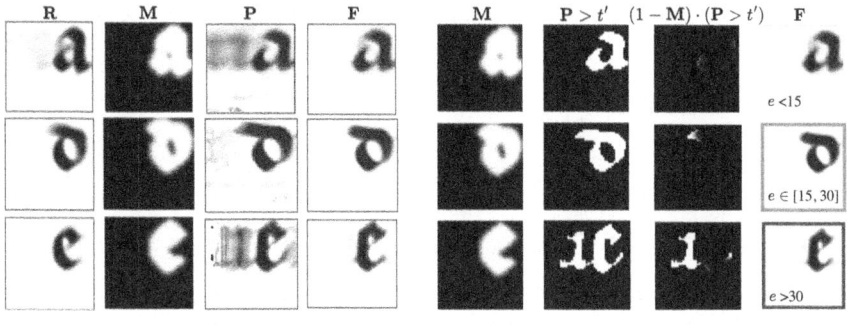

(a) **Filtering strategy** (b) **Failure case identification**

Fig. 2. Prototype filtering and failure case identification. We use a mask M defined from the reference prototype **R** to remove artifacts from finetuned prototypes **P**, yielding a filtered prototype **F**. We compute an error e associated to the filtering to automatically identify potential failure cases. (Color figure online)

Character Prototype Filtering. While the reference prototypes generally are of high quality, we observed various artifacts in the finetuned prototypes, particularly for less common characters, when trained on a single document. While this does not hinder the qualitative analysis of the prototypes' shapes, it does complicate the quantitative comparison between prototypes. To alleviate this issue, we propose to filter the finetuned prototypes using the reference ones, as visualized in Fig. 2.

Let us consider a specific character, the associated reference prototype **R** and the associated finetuned prototype **P** for a given finetuning. Using the reference prototype, we define a reference mask **M** as

$$\mathbf{M} = \mathbf{G} * D(\mathbf{R} > t), \qquad (1)$$

where **G** is a Gaussian filter, $*$ denotes a convolution, D is a dilation operation and $\mathbf{R} > t$ is the binary mask associated to pixels for which **R** is greater than a threshold t. In our experiments, we use a Gaussian **G** of standard deviation 2, a dilation D of 2 pixels, and a threshold $t = 0.8$. Intuitively, this mask defines in a soft way, for each character, pixels that are close to the reference prototype.

Using this mask, we define a filtered prototype $\mathbf{F} = \mathbf{M} \cdot \mathbf{P}$, where $\cdot$ is the pixel-wise multiplication, which we use for all of our analyses.

Automatic Identification of Failure Cases. While the filtering process described above generally improves the visual quality of the prototypes without changing the appearance of the characters themselves, there are instances where either the appearance is slightly altered or the finetuned prototype is of very low quality. We want to identify such cases automatically, to avoid misinterpretations. To do so, for a given character associated with a reference mask **M** and a finetuned prototype **P** we compute the error e defined by:

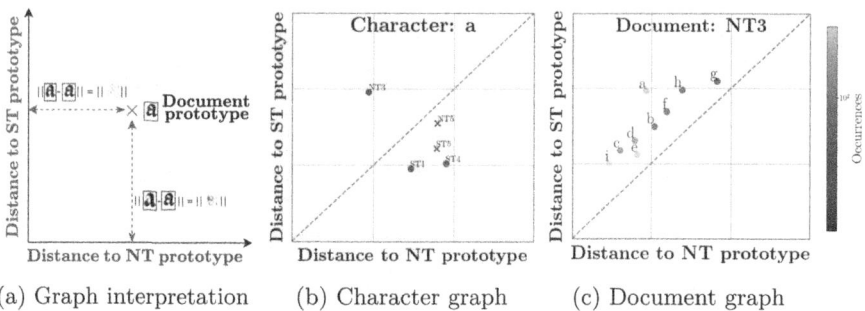

Fig. 3. Comparison graphs. The markers correspond to different document prototypes and their coordinates to their distance to the Northern and Southern *Textualis* prototypes. See text for details. (Color figure online)

$$e = \|(1 - \mathbf{M}) \cdot (\mathbf{P} > t')\|, \qquad (2)$$

where $\| \ \|$ is the norm of an image, $\cdot$ is the pixel-wise multiplication, and t' is a scalar threshold, set to 0.65 in our experiments. Intuitively, this error can be interpreted as the number of pixels that are present in the finetuned prototype $\mathbf{P}$ (i.e., have values higher than t') but are filtered out by the mask M. This value e enables us to identify: (i) finetuned prototypes whose shape is significantly different from the reference one and are thus modified by the filtering process, such as the ‹d› in Fig. 2, and (ii) finetuned prototypes of low quality that might not be easily interpretable, such as the ‹e› in Fig. 2. In our results, we highlight prototypes where $e > 15$ in orange and prototypes where $e > 30$ in red.

3.2 Character Prototype Comparison for Palaeographic Analysis

Visual Comparison. We can visually highlight the morphological differences between two prototypes by subtracting one from the other. To make this difference easier to understand, we use a colormap that represents zeros as white, and positive and negative values as two distinct colors, typically red and blue. This method reveals pixel-wise differences, facilitating an initial qualitative examination of the morphological disparities (see Table 2 and Fig. 7).

Character and Document Comparison Graphs. To quantitatively analyze character prototypes, we introduce an adapted comparison graph, illustrated in Fig. 3. In this graph, each point represents a specific document character prototype, with its coordinates defined as its distance in pixel space to two selected prototypes. Since we study *Textualis Formata*, we use the prototypes of Northern and Southern *Textualis* (NT and ST) (see Sect. 4.1 for details), other prototypes could be selected for different analysis. The distance to the axes can be interpreted as visualized in Fig. 3a.

We employ two complementary types of graphs for our analysis. Firstly, *character graphs* (see Fig. 3b) concentrate on a single character across all documents. In these graphs, dots represent the documents selected to train the reference, Northern and Southern *Textualis* models, while crosses denote the remaining documents. Blue, resp. red, markers signify Northern, resp. Southern *Textualis* documents. The identifier for each document is written near its marker. This type of graph allows to easily identify outlier documents for a specific character, such as NT5 for the ⟨a⟩ character, which is closer to the ST prototype. Secondly, *document graphs* (see Fig. 3c) focus on a specific document, where each dot corresponds to a different character, labeled near the marker. The color of the dots corresponds to the frequency of occurrence, with darker dots representing less frequent characters. This type of graph facilitates the identification, for a given document, of the characters most typical of a subtype, such as the ⟨a⟩ for NT3.

Quantifying Character Variability. According to the literature, the Northern *Textualis* class allows for more morphological variation across documents than the Southern *Textualis*. While the visualizations and graphs can qualitatively support this idea, we further aim to quantify the characters' variability. Thus, we report the standard deviations of character prototypes within one subtype, σ_{NT} for Northern *Textualis* and σ_{ST} for Southern *Textualis*. These standard deviations can be thought of as the average number of pixels that change across two character prototypes of the same subtype.

4 Experiments

Research Question and Analysis Framework. To analyze the results of our approach, we adopt the taxonomy formalized by A. Derolez [16], which provides a framework based on morphological criteria. We select a corpus in *Textualis* script type, specifically in its canonized calligraphic form *Formata*, due to its more distinguishable morphological elements compared to more rapidly executed forms. Despite the morphology-based categorization where *Textualis Formata* represents a coherent group, Derolez makes "an important distinction between two fundamentally different species", Northern and Southern *Textualis*, based on their geographical distribution and a set of minute morphological differences. However, these differences vary according to factors such as century, geographical origin, or language, and often intersect, blurring this fundamental distinction in some cases. Our goal is to confront the results of our approach with Derolez's criteria and observations.

Table 1. Dataset Description. The 'Doc.' column refers to the names we use for the different documents in this paper, NT and ST stand for Northern and Southern *Textualis*, and the 'Ref.' column reports which documents are used to train the reference and subtype models.

Doc.	Ref.	Shelfmark	Language	Century	Date	Origin	Folio(s)	Lines
NT1	✓	Paris, BnF, Français 403 [62]	French	13th	1226-1250	England	4r	76
NT2	✓	Paris, BnF, Français 12400 [62]	French	14th	1305-1310	Eastern France	92r	54
NT3	✓	Arras, BM Ms. 861 (315) [12]	Latin	14th	-	-	56r	65
NT4	✓	Paris, BnF, Français 1728 [44]	French	14th	1372	-	3r	59
NT5		Paris, BnF, Français 20120 [62]	French	13th	1240-1250	Paris or Orleans	7r	81
NT6		Paris, BnF, Français 619 [35]	French	14th	1375-1400	-	1v	65
NT7		Berlin, SB, Hdschr. 25 [12]	Latin	15th	1451-1500	Flanders	22r-22v	25
ST1	✓	Paris, BnF, Français 9082 [62]	French	13th	1295	Rome	171r	56
ST2	✓	Paris, BnF, Espagnol 65 [5]	Navarrese	14th	1301-1310	-	5v,6r	118
ST3	✓	Paris, BnF, Italien 590 [2]	Italian	14th	1370-1410	Italy	18r	68
ST4	✓	Madrid, FLG, mss 289 (Hand A) [22]	Castilian	15th	1480	Seville	245v	86
ST5		Paris, BnF, Français 187 [62]	French	14th	1350-1386	Milan or Genova	18r	16
ST6		Paris, BnF, Latin 7720 [23]	Latin	15th	1390-1410	Florence	102v	74
ST7		Madrid, FLG, mss 289 (Hand B) [22]	Castilian	15th	1480	Seville	274v	52
Total:								892

4.1 Dataset and Experiment Details

Data Selection. Our data was build from two open-access repositories, ECMEN [62,63] and CATMuS [45]. We selected the documents to ensure the variability of the corpus in terms of geographic, linguistic, and chronological distribution, resulting in seven documents for each subscript as listed in Table 1 (labelled NT 1-7 and ST 1-7). We verified and normalized the transcriptions to fit a graphemic approach [12,13]. Despite our efforts to use diverse and representative documents, we acknowledge that biases may exist within our dataset.

Character Set Choice. From the extended set of characters in medieval manuscripts - including upper and lower case letters, ligatures, punctuation, and abbreviation signs - we follow a standard approach for morphological analysis and focus on the lowercase alphabetic characters where morphology is crystalized through frequent usage. From these characters, we show results on the ones common to all documents, thus excluding ⟨j,k,v,x,y,z⟩.

Trained Models. For our analysis, we trained multiple models to obtain character prototypes at different levels of granularity: (i) a script type model for *Textualis*, (ii) script subtype models for Northern and Southern *Textualis*, and (iii) document level models for each document in our dataset. We use the *Textualis* script type model as reference model, and finetune all other models from it as explained in Sect. 3.1. To validate that our reference and subtype models can be effective to analyze documents they were not trained on, we limited their training to NT 1-4 and ST 1-4, indicated in the 'Ref' column in Table 1.

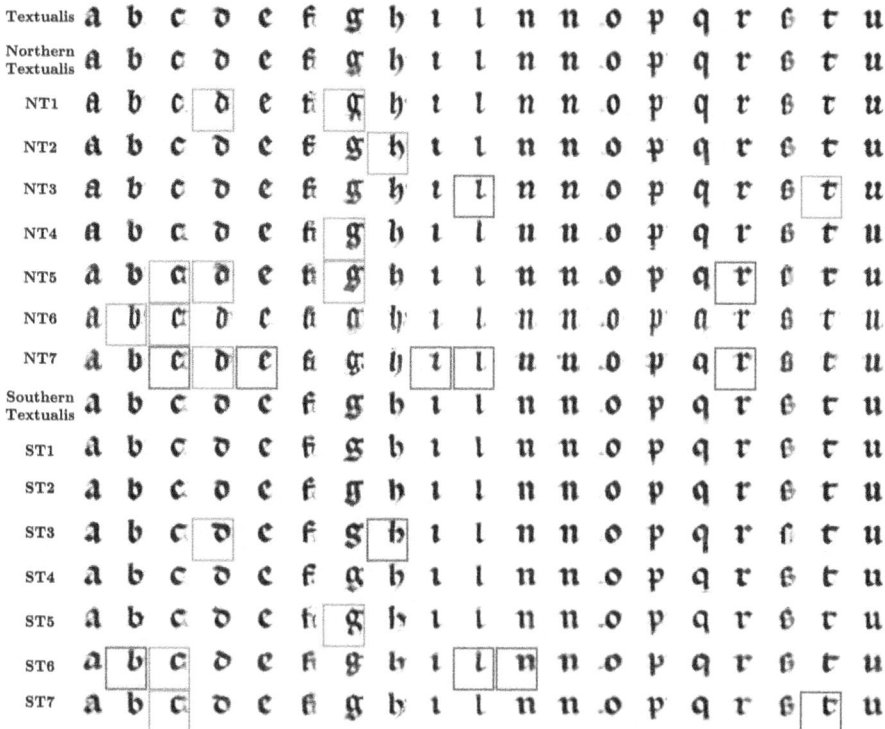

Fig. 4. Our filtered prototypes on type, sub-type and document level. The highlighted prototypes are the ones for which filtering had a significant impact (see Sects. 3.1 and 4.1 for details). (Color figure online)

4.2 General Results

Prototype Quality. Figure 4 shows the character prototypes generated by our approach across type, sub-type, and document levels. We note several limits in these prototypes. Firstly, the prototype for ⟨m⟩ is not well modeled (𝐧). This issue stems from the network being trained with a CTC loss allowing to use the same prototype twice to model the same letter. As a result, we exclude ⟨m⟩ from our analysis. Secondly, the two allographs ⟨ſ/s⟩ are represented by a single prototype resulting in an averaged representation of the two (ſ), necessitating cautious examination. Thirdly, we meticulously scrutinized all prototypes highlighted in orange and red, where filtering significantly impacted the outcomes. In almost all instances, the filtering was meaningful, eliminating irrelevant artifacts from the prototypes. Nevertheless, it is worth mentioning that: (i) the lengthy shaft of the ST3 ⟨d⟩ (ð) and the hairline extension of the limb of ⟨h⟩ in NT2 (h) are slightly severed; (ii) in general, documents that are not part of the reference set, are less accurately modeled and more impacted by the filtering, especially the forked ascender tops of ⟨h⟩ and ⟨b⟩. This information loss, while limited, should be considered during the analysis.

Palaeographical Relevance of the Subtype Prototypes. To showcase how our prototypes can be related to classical palaeographic analysis, we systematically compare in Table 2 Derolez's general morphological criteria to our Northern and Southern *Textualis* prototypes, highlighting their variations by visualizing their difference. We find that Derolez's observations closely align with the variations that our prototypes enable us to visualize. Additionally, we report our variability scores σ_{NT} and σ_{ST} for each letter, which were consistently higher for Northern *Textualis*, which is consistent with Derolez's claim that this script subtype generally exhibits higher intra-class variation.

4.3 Character Graph Analysis

In this section, we provide examples of how our character graphs, together with our prototypes, can support a detailed palaeographic analysis of the variations of a specific character (examples presented in Fig. 5).

Discriminative Characters. We first analyze the results for four discriminative characters, ⟨a,o,p,h⟩. The letter ⟨a⟩ is often considered as a distinguishing criterion between script types, so much so that W. Oeser [41] distinguished seven categories within the Northern *Textualis* script subtype mainly based on allographs of ⟨a⟩. Most striking in our ⟨a⟩ character graph is that the prototypes for NT5 (𝖆) and NT7 (𝖆) are actually closer to the ST prototypes. This is consistent with the observation that open ⟨a⟩ forms are standard for ST. The dispersion of the characters on the graph also provides insight into the variability of ⟨a⟩ in this subtype. The group associated to NT1-4 corresponds to the closed "box-a" form in NT2 (𝖆) and NT4 (𝖆) and the double-bow variant in NT1 (𝖆) and NT3 (𝖆). NT6 presents a more vertically elongated form and stands out (𝖆). While there is morphological variations across ST documents, with round shapes (ST1 𝖆; ST2 𝖆; ST5 𝖆; ST6 𝖆), or with more angular inner bows (ST3 𝖆; ST4 𝖆; ST7 𝖆), the consistent use of an open form, or only closed with a hairline, distinguishes them from the NT subtype, and all ST documents prototypes are closer to the ST prototype.

The letter ⟨o⟩ is also particularly discriminative. The treatment of its (generally) two mirroring arcs, using broken or semi-circular strokes, often has visual echoes in letters with lobes and arcs like ⟨b, c, e, p, q⟩, which contributes to the visual evaluation of a hand or script type as wide/round or narrow/angular. This is confirmed by the fact that all the points of documents identified as NT are above the diagonal and all the ones corresponding to ST below, meaning that the prototypes for each document are closer to its subtype prototype than the other one. For NT, the forms of NT2-5 (𝖔𝖔𝖔𝖔) are particularly well reconstructed, better than NT1 (𝖔) and NT7 (𝖔) which are slightly more narrow and vertically elongated. Again, the form of NT6 stands out (𝖔), which consists of double broken strokes resulting in a narrow, quadrangle shape. Similarly for

Table 2. Comparison between Derolez' criteria for Northern and Southern *Textualis* and our subtype prototypes.

⟨Ch.⟩	Derolez'criteria	NT σ_{NT}	ST σ_{ST}	diff.
⟨a⟩	**NT**: Closed form with variations like "box-⟨a⟩" **ST**: Open form or slightly closed with hairline	4.0	3.4	
⟨b⟩	**NT**: Sloped or forked ascender tops **ST**: (i) Flat ascender tops, (ii) round lobe	4.1	3.6	
⟨c⟩	**NT**: Angular or broken lobe curves **ST**: Semi-circular lobe	2.9	2.4	
⟨d⟩	**NT**: (i) Lengthened and (ii) concave shaft **ST**: (i) Shorter shaft and (ii) almost horizontal, (iii) round bowl	3.8	3.1	
⟨e⟩	**NT**: (i) Diagonal direction of the hairline and (ii) angular or broken lobe curves **ST**: (i) Horizontal or no hairline, (ii) semi-circular lobe form	3.3	3.2	
⟨f⟩	**NT**: Incurvation of the shaft foot to the right **ST**: Flat foot	4.3	4.3	
⟨g⟩	**NT**: Tendency for the closed, "8-shaped" form **ST**: Tendency for the open, "Rücken -g" form **Note**: Various intermediate forms and difficult to classify	5.8	5.4	
⟨h⟩	**NT**: (i) Incurvation of the shaft foot to the right, (ii) extended or dislocated limb and (iii) sloped or forked tops **ST**: (i) Flat ascender foot, (ii) circular limb on the baseline and (iii) flat ascender tops	4.9	4.2	
⟨i⟩	**NT**: (i) Accentuated (diamond-shaped or forked) headline and (ii) extended hairline for the foot **ST**: (i) Approach stroke for the headline and (ii) flat end for the foot	2.4	1.6	
⟨l⟩	**NT**: Sloped or forked ascender tops **ST**: Flat tops	2.5	1.8	
⟨n⟩	**NT**: Accentuated (diamond-shaped or hairlines) for the headline and (ii) same for feet **ST**: (i) Approach stroke hairline for the headline and (ii) flat feet	3.7	3.0	
⟨o⟩	**NT**: Broken / more vertically elongated curves **ST**: Circular arc forms	3.5	2.7	
⟨p⟩	**NT**: (i) Artificial spurs on the left and (i) decorated descender feet **ST**: No spurs and (ii) flat descender feet	4.2	3.2	
⟨q⟩	**NT**: (i) Lengthy and (ii) decorated descenders **ST**: (i) Short and (ii) flat descenders	4.8	4.0	
⟨r⟩	**NT**: (i) Hairline endstroke for shaft foot and (ii) angular horizontal stroke **ST**: (i) Flat shaft foot and (ii) straight horizontal stroke	2.8	2.4	
⟨s⟩	**NT**: Incurvation of the shaft foot to the right for ⟨ſ⟩ and (ii) closed and angular curves for ⟨s⟩ **ST**: (i) Flat shaft foot for ⟨ſ⟩ and (ii) open semi-circular curves for ⟨s⟩	2.7	2.4	
⟨t⟩	**NT**: Vertical pendant hairline of the headstroke **ST**: No ornaments **Note**: Different levels of shaft projection above headline and length of horizontal stroke	3.0	2.4	
⟨u⟩	**NT**: Accentuated (diamond-shaped or sloped) headline **ST**: Flat or left approach stroke for headline	3.7	3.0	

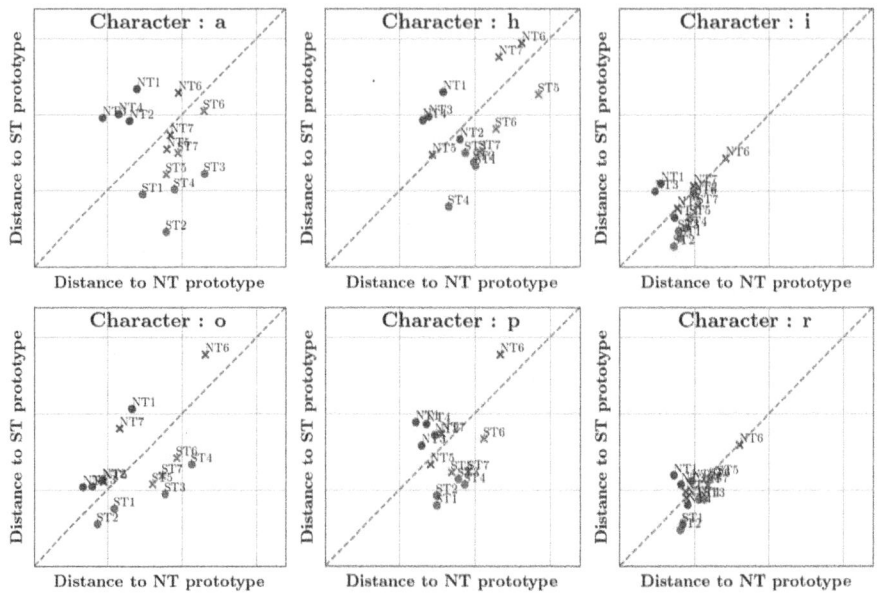

Fig. 5. Character comparison graphs. (Color figure online)

ST, ST1 (o) and ST2 (o) are particularly close to the ST prototype, being less wide than ST3-7 (ooooo). This is consistent with Derolez' assertion that angularity/roundness separates the two subtypes, while intra-class variation is associated with different degrees of narrowness/breadth.

Regarding the letter ⟨h⟩, it highlights one of the limitations of our approach. The extended limb, characteristic of NT (cf. Table 2) is clearly present in all associated documents and prototypes. However, because its position varies in a document, the limb appears dimmed in the NT prototype (h). Moreover, its position varies significantly across different documents, which can result in a greater distance between a document prototype and the NT prototype. This explains why the NT2 prototype is actually more similar to the ST prototype, while NT5 is as close to both. Note that the shift of the limb in NT2 (not curved to the left but rather extended straight down) compared to the NT prototype is so significant that our filtering partially erases it (as can be confirmed by examining the masked region, similar to Fig. 2b), which was flagged by our automatic failure identification. This emphasizes the necessity to confront our graphs with the visual appearance of the prototypes and the documents for interpretation.

For the letter ⟨p⟩, we note that the consistent presence of artificial spurs at the baseline level and in general of decorations are characteristic of NT (p). The two subtypes appear well separated, except for the deviation of NT5 (p), which we will further analyze in Sect. 4.4.

Non-Discriminative Characters. We now examine ‹i› and ‹r›. In our graphs, the points corresponding to these letters in all documents are close to the diagonal, i.e., they are as close to both the NT and ST prototypes. The fact that they are almost all close to the origin indicates that they present little variation. For ‹i›, NT1 and NT3 stand out as more typical of NT, and they indeed show clear diamond-shaped headlines (**ı ı**). The NT4 prototype (**ı**), on the other hand, is actually closer to the ST prototypes, and it does not present any headline decoration typical of NT. For both characters, the NT6 prototypes are much further than the rest from both the NT and ST prototypes and correspond to much narrower and elongated forms (**ı r**).

4.4 Document Graph Analysis

In this section, we discuss and interpret document graphs for the examples presented in Fig. 6. Additionally, we visualize the differences between the document prototypes and subtype prototypes in Fig. 7, leveraging them to better understand the graphs.

Class-Representative Cases. We start by examining two documents that are very typical of their subtype. For **NT3**, a 14th century manuscript in Latin, the graph clearly shows that all the character prototypes are closer to the NT prototypes than the ST prototypes. More in details, NT3 presents a closed, double bow ‹a› (**a**), which progressively dominated over other variants from the end of the 13th c. [16], ascenders are consistently sloped on the left side (**b h l**, hairlines directed to the right at the baselines (**f i l n e t**), and there are clear diamond-shaped headlines (**ı u**). All characters thus fully correspond to the expected NT forms. For **ST2**, copied in the first decade of 14th c. in the Iberian peninsula, the character prototypes are, on the contrary, all closer and conforming to the ST prototypes: compact letters with very short flat-top ascenders (**b h**), flat-feet descenders (**p q**) and strokes so bold, hairlines almost become invisible (**c b**).

Ambiguous Cases. A particular interest of our document graph method is the identification and analysis of documents that partially diverge from their assigned subtype. **NT5** stands out in the graphs, as seven character prototypes are closer to the ST than NT prototypes, with a particular difference for ‹a,g,q› (**a g q**). Copied in the 13th c., in the Paris/Orleans area, NT5's forms are significantly smaller in size, an example of "pearl-script", generally used for Parisian pocket-Bibles. Even though its level of execution is still *Formata*, due to their size, certain letters are simplified, resulting in forms that are closer to ST, like open ‹a› (**a**) - a characteristic of early NT samples -, and spurless ‹p› (**p**). On the contrary, a closer examination of ‹g› reveals it is in reality characteristic of NT, but not well modeled by our prototypes, in part because of the high level of variation, resulting in a blurred prototype, and in part because it was too different from the reference documents, which was actually flagged by our automatic failure case identification. **ST6**, copied in Florence at the end of 14th/beginning of 15th c., also presents two class diverging characters, ‹b›

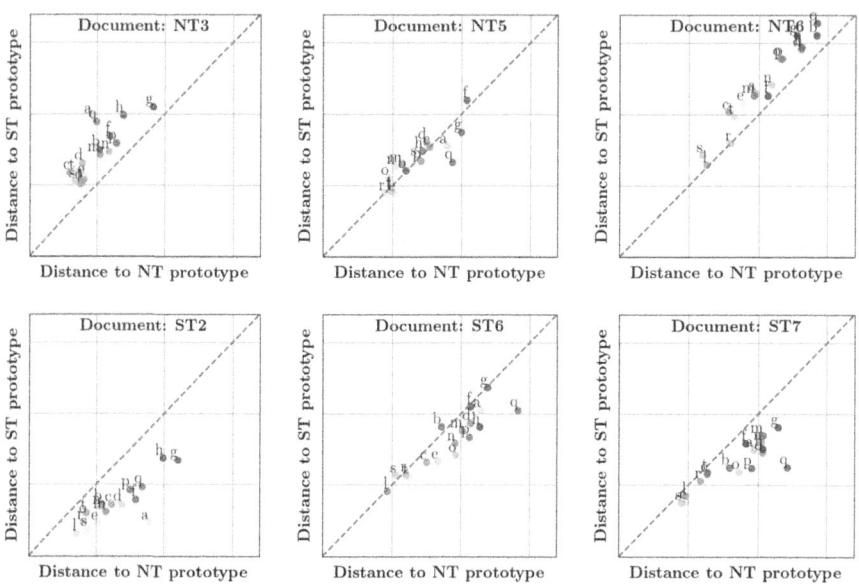

Fig. 6. Document comparison graphs. (Color figure online)

and ⟨s/ſ⟩. For ⟨s/ſ⟩, even though not directly obvious due to the fusion of the two allographs, ⟨ſ⟩'s foot is curved to the right (ſ), a characteristic proper to NT. At the same time, ⟨b⟩'s slightly dislocated lobe (b), also present in ⟨h⟩ (h) and ⟨p⟩ (p), characteristic of later examples of *Textualis* [16], is disrupting the typical circular lobe shape of ST. These dislocated lobes are clear in our Fig. 7.

Later Textualis Examples. Later specimens of both subtypes present notable differences from their earlier counterparts. **NT6**, copied in the last quarter of the 14th c. is an example of later Northern *Textualis*, and, while the graph shows that all prototypes are closer to the NT prototypes than the ST prototypes, it also clearly shows that they are very different from both, i.e., they are more on the top right of the graph. This document particularity lies in its strict angular forms, with diamond-shaped minim feet (1 n) as well as its exaggerated narrow shapes, with a total absence of round strokes for arcs (o c o), typical of 14th-15th c. (esp. Northern) *Textualis*. **ST7** is an example of late, 15th c. Iberian *Textualis*, discussed separately by Derolez due to its tendency towards more angular forms. Particularities of this type lie in the presence of hairlines and angular shapes (a c c) alongside typical rounder ones (o o q) with flat tops and feet (b b n ß). However, these characteristics are only discernible while looking at the prototypes and their differences, and not directly in the graph, where the prototypes are closer to ST, and not particularly poorly reconstructed. This can be understood both by the fact that these particularities do not make the prototypes more similar to the NT ones, and by the fact that another Iberian *Textualis* (ST4), from the same manuscript but from a different hand, was used in

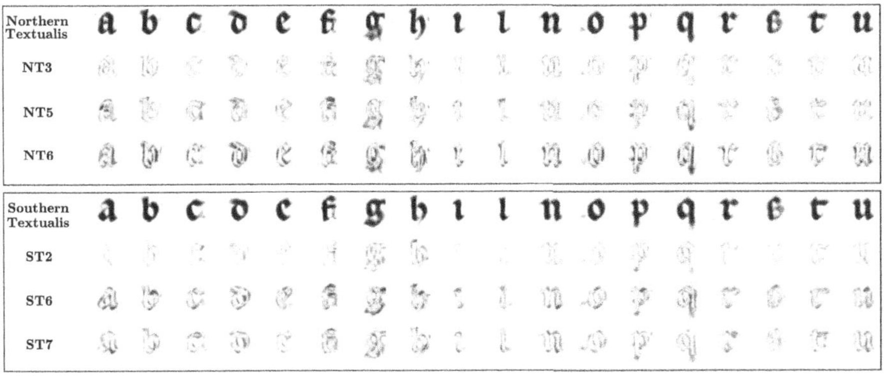

	a	b	c	d	e	f	g	h	i	l	n	o	p	q	r	s	t	u
Northern Textualis	a	b	c	d	e	f	g	h	i	l	n	o	p	q	r	s	t	u
NT3																		
NT5																		
NT6																		
Southern Textualis	a	b	c	d	e	f	g	h	i	l	n	o	p	q	r	s	t	u
ST2																		
ST6																		
ST7																		

Fig. 7. Visual comparison. Subtype and document prototypes and their pixel-wise differences with positive values in blue and negative in red. (Color figure online)

our reference set, and thus our ST prototype already models some characteristics of this *Textualis* type. This highlights the impact of the prototype training data on our analysis, prototypes utilized as the basis for our comparison graph axes.

5 Conclusion

In this work, we introduced a deep learning-based methodology for interpretable script comparison and analysis. By applying it to the two subtypes of *Textualis Formata* script type defined by A. Derolez -Northern and Southern *Textualis*- we showed how such an approach can complement qualitative document analysis, by quantifying specific elements and summarizing information. We believe our approach contributes to bridging the gap between traditional and learning-based approaches to paleography.

Aknowledgements. This study was supported by the CNRS through MITI and the 80|Prime program (CrEMe Caractérisation des écritures médiévales), and by the European Research Council (ERC project DISCOVER, number 101076028). We thank Ségolène Albouy, Raphaël Baena, Sonat Baltacı, Syrine Kalleli, and Elliot Vincent for valuable feedback.

References

1. Aiolli, F., Simi, M., Sona, D., Sperduti, A., Starita, A., Zaccagnini, G.: SPI: a system for palaeographic inspections. AIIA Notizie **4**, 34–38 (1999)
2. Alba, R., Rubin, G., Boschetti, F., Fischer, F., Clérice, T., Chagué, A.: HTRomance, medieval Italian corpus of ground-truth for handwritten text recognition and layout segmentation [dataset] (2023). https://doi.org/10.5281/zenodo.8272751, https://github.com/HTRomance-Project/medieval-italian, v1.0.1

3. Baird, H.S.: Model-directed document image analysis. In: Proceedings of the Symposium on Document Image Understanding Technology, vol. 1 (1999)
4. Berg-Kirkpatrick, T., Durrett, G., Klein, D.: Unsupervised transcription of historical documents. In: Proceedings of the 51st Annual Meeting of the Association for Computational Linguistics (Volume 1: Long Papers), pp. 207–217 (2013)
5. Bordier, J., Gille Levenson, M., Brisville-Fertin, O., Clérice, T., Chagué, A.: HTRomance, medieval Spain corpus of ground-truth for Handwritten text recognition and layout segmentation [dataset] (2023). https://github.com/HTRomance-Project/middle-ages-in-spain, v0.0.6
6. Cencetti, G.: Lineamenti di Storia della scrittura latina: dalle lezioni di Paleografia (Bologna a.a. 1953-54). Guerrini Ferri, G., Bologna (1997)
7. Christlein, V., Bernecker, D., Maier, A., Angelopoulou, E.: Offline writer identification using convolutional neural network activation features. In: Gall, J., Gehler, P., Leibe, B. (eds.) GCPR 2015. LNCS, vol. 9358, pp. 540–552. Springer, Cham (2015). https://doi.org/10.1007/978-3-319-24947-6_45
8. Ciula, A.: Digital palaeography: using the digital representation of medieval script to support palaeographic analysis. Digital Medievalist **1** (2005)
9. Cloppet, F., et al.: New Tools for exploring, analysing and categorising medieval scripts. Digital Medievalist **7** (2012). https://doi.org/10.16995/dm.44
10. Cloppet, F., Eglin, V., Helias-Baron, M., Kieu, V.C., Stutzmann, D., Vincent, N.: ICDAR 2017 competition on the classification of medieval handwritings in Latin script. In: 14th IAPR International Conference on Document Analysis and Recognition. ICDAR 2017, pp. 1371–1376. CPS, Kyoto (2017). https://doi.org/10.1109/ICDAR.2017.224
11. Cloppet, F., Eglin, V., Kieu, V.C., Stutzmann, D., Vincent, N.: ICFHR2016 competition on the classification of medieval handwritings in Latin script. Proceedings of International Conference on Frontiers in Handwriting Recognition, pp. 590–595 (2016)
12. Clérice, T., Chagué, A., Vlachou-Efstathiou, M.: CREMMA Medii Aevi [dataset] (2023). https://github.com/HTR-United/CREMMA-Medieval-LAT, v0.1.2
13. Clérice, T., Pinche, A.: Choco-Mufin, a tool for controlling characters used in OCR and HTR projects (2021). https://doi.org/10.5281/zenodo.5356154, https://github.com/PonteIneptique/choco-mufin
14. Daher, H., Églin, V., Brès, S., Vincent, N.: Étude de la dynamique des écritures médiévales: analyse et classification des formes écrites. Gazette du livre médiéval **56**(1), 21–41 (2011)
15. Davis, L.F.: Towards an automated system of script classification. Manuscripta **42**(3), 193–201 (1998)
16. Derolez, A.: The Palaeography of Gothic Manuscript Books: From the Twelfth to the Early Sixteenth Century. Cambridge University Press (2003)
17. Djeddi, C., Meslati, L.S., Siddiqi, I., Ennaji, A., El Abed, H., Gattal, A.: Evaluation of texture features for offline Arabic writer identification. In: 2014 11th IAPR International Workshop on Document Analysis Systems, pp. 106–110. IEEE (2014)
18. Djeddi, C., Siddiqi, I., Souici-Meslati, L., Ennaji, A.: Codebook for writer characterization: a vocabulary of patterns or a mere representation space? In: 2013 12th International Conference on Document Analysis and Recognition, pp. 423–427. IEEE (2013)
19. Gasparri, F.: Remarques sur la terminologie paléographique. Revue d'Histoire des Textes **13**(1964), 111–114 (1966)

20. Gilissen, L.: L'expertise des écritures médiévales: recherche d'une méthode avec application à un manuscrit du XIe siècle: le lectionnaire de Lobbes, Codex Bruxellensis 18018. Scriptorium/Les Publications de Scriptorium **6** (1973)
21. Gilissen, L.: III. ductus et rapport modulaire. Scriptorium **29**(2), 235–244 (1975)
22. Gille Levenson, M.: Towards a general open dataset and model for late medieval Castilian text recognition (HTR/OCR). J. Data Min. Digit. Humanit. (2023). https://doi.org/10.46298/jdmdh.10416
23. Glaise, A., Clérice, T., Boschetti, F., Fischer, F., Chagué, A.: HTRomance, medieval Latin corpus of ground-truth for handwritten text recognition and layout segmentation [dataset] (2024). https://github.com/HTRomance-Project/medieval-latin, v0.0.6
24. Goyal, K., Dyer, C., Warren, C., G'Sell, M., Berg-Kirkpatrick, T.: A probabilistic generative model for typographical analysis of early modern printing. arXiv preprint arXiv:2005.01646 (2020)
25. Hannad, Y., Siddiqi, I., El Kettani, M.E.Y.: Writer identification using texture descriptors of handwritten fragments. Expert Syst. Appl. **47**, 14–22 (2016)
26. Hassner, T., Rehbein, M., Stokes, P.A., Wolf, L.: Computation and palaeography: potentials and limits. Kodikologie und Paläographie im digitalen Zeitalter **3**, 1–30 (2015)
27. He, S., Schomaker, L.: Delta-n hinge: rotation-invariant features for writer identification. In: 2014 22nd International Conference on Pattern Recognition, pp. 2023–2028. IEEE (2014)
28. He, S., Schomaker, L.: Deep adaptive learning for writer identification based on single handwritten word images. Pattern Recogn. **88**, 64–74 (2019)
29. He, S., Wiering, M., Schomaker, L.: Junction detection in handwritten documents and its application to writer identification. Pattern Recogn. **48**(12), 4036–4048 (2015)
30. Hochberg, J., Kelly, P., Thomas, T., Kerns, L.: Automatic script identification from document images using cluster-based templates. IEEE Trans. Pattern Anal. Mach. Intell. **19**(2), 176–181 (1997)
31. Kestemont, M., Christlein, V., Stutzmann, D.: Artificial paleography: computational approaches to identifying script types in medieval manuscripts. Speculum **92**(S1), S86–S109 (2017)
32. Kopec, G.E., Lomelin, M.: Supervised template estimation for document image decoding. IEEE Trans. Pattern Anal. Mach. Intell. **19**(12), 1313–1324 (1997)
33. Kordon, F., et al.: Classification of incunable glyphs and out-of-distribution detection with joint energy-based models. Int. J. Doc. Anal. Recogn. (IJDAR) **26**(3), 223–240 (2023)
34. Lebourgeois, F., Moalla, I.: Caractérisation des écritures médiévales par des méthodes statistiques basées sur les cooccurrences. Gazette du livre médiéval **56-57**, 72–100 (2011)
35. Leroy, N., Pinche, A., Camps, J.B., Clérice, T., Chagué, A.: HTRomance, medieval French corpus of ground-truth for handwritten text recognition and layout segmentation [dataset]. https://github.com/HTRomance-Project/medieval-french, v0.0.7
36. Mamatsis, A.R., Mamatsi, E., Chalatsis, C., Arabadjis, D., Kampouri, P., Papaodysseus, C.: A novel methodology for writer (hand) identification: establishing rigas feraios wrote two important Greek documents discovered in Romania. Heritage Sci. **11**(1), 38 (2023)
37. McGillivray, M.: Statistical analysis of digital paleographic data: what can it tell us? Digital Studies/Le champ numérique **11** (2005)

38. Moalla, I., Lebourgeois, F., Emptoz, H., Alimi, A.: Image analysis for palaeography inspection. In: Second International Conference on Document Image Analysis for Libraries (DIAL2006), pp. 8–311. IEEE (2006)
39. Muzerelle, D.: À la recherche d'algorithmes experts en écritures médiévales. Gazette du livre médiéval **56**(1), 5–20 (2011). https://doi.org/10.3406/galim.2011.1979
40. Nigam, S., Verma, S., Nagabhushan, P.: Document analysis and recognition: a survey. Authorea Preprints (2021)
41. Oeser, W.: Das «a» als Grundlage für Schriftvarianten in der gotischen Buchschrift. Scriptorium **25**(1), 25–45 (1971)
42. Ornato, E.: Ii. statistique et paléographie: peut-on utiliser le rapport modulaire dans l'expertise des écritures médiévales? Scriptorium **29**(2), 198–234 (1975)
43. Parkes, M.B.: English Cursive Book Hands, 1250–1500. Clarendon, Oxford (1969)
44. Pinche, A.: Cremma Medieval (2023) [dataset]. https://github.com/HTR-United/cremma-medieval
45. Pinche, A., et al.: Catmus-medieval: consistent approaches to transcribing manuscripts (2023). https://univ-lyon3.hal.science/hal-04453952v1
46. Poulle, E.: Paléographie et méthodologie: vers l'analyse scientifique des écritures médiévales. Bibliothèque de l'École des chartes **132**(1), 101–110 (1974)
47. Ramel, J.Y., Sidère, N., Rayar, F.: Interactive layout analysis, content extraction, and transcription of historical printed books using pattern redundancy analysis. Literary Linguist. Comput. **28**(2), 301–314 (2013)
48. Schomaker, L., Bulacu, M.: Automatic writer identification using connected-component contours and edge-based features of uppercase western script. IEEE Trans. Pattern Anal. Mach. Intell. **26**(6), 787–798 (2004)
49. Schomaker, L., Franke, K., Bulacu, M.: Using codebooks of fragmented connected-component contours in forensic and historic writer identification. Pattern Recogn. Lett. **28**(6), 719–727 (2007)
50. Siddiqi, I., Vincent, N.: Text independent writer recognition using redundant writing patterns with contour-based orientation and curvature features. Pattern Recogn. **43**(11), 3853–3865 (2010)
51. Siglidis, I., Gonthier, N., Gaubil, J., Monnier, T., Aubry, M.: The learnable typewriter: a generative approach to text line analysis (2023). https://arxiv.org/abs/2302.01660
52. Sirat, C.: L'examen des écritures: l'œil et la machine: essai de méthodologie. Ed. du Centre National de la Recherche Scientifique (1981)
53. Smith, M.: (review) derolez (albert), the palaeography of gothic manuscript books. from the twelfth to the early sixteenth century, cambridge, 2003. Scriptorium **58**(2), 274–279 (2004)
54. Sommerschield, T., et al.: Machine learning for ancient languages: a survey. Comput. Linguist. **49**(3), 703–747 (2023)
55. Stansbury, M.: The computer and the classification of script. In: Kodikologie und Paläographie im digitalen Zeitalter - Codicology and Palaeography in the Digital Age, vol. 2, p. 238. BoD, Norderstedt (2009)
56. Stokes, P.A.: Describing handwriting, part i-v. Blog Post (2011). https://digipal.eu/blog/describing-handwriting-part-i/. Accessed 15 Mar 2024
57. Stutzmann, D.: Variability as a key factor for understanding medieval scripts: the oriflamms project (anr-12-corp-0010). In: Brookes, S., Rehbein, M., Stokes, P. (eds.) Digital Palaeography. Digital Research in the Arts and Humanities, Routledge. https://halshs.archives-ouvertes.fr/halshs-01778620

58. Stutzmann, D.: Nomenklatur der gotischen Buchschriften: Nennen? Systematisieren? Wie und wozu? (Rezension über: Albert Derolez: The Palaeography of Gothic Manuscript Books. From the Twelfth to the Early Sixteenth Century. Cambridge u.a.: Cambridge University Press 2003). IASLonline (2005). http://www.iaslonline.de/index.php?vorgang_id=995
59. Stutzmann, D.: Paléographie statistique pour décrire, identifier, dater... Normaliser pour coopérer et aller plus loin ? In: Kodikologie und Paläographie im digitalen Zeitalter 2 - Codicology and Palaeography in the Digital Age 2, pp. 247–277. No. 3 in Schriften des Instituts für Dokumentologie und Editorik, BoD, Norderstedt (2010). https://kups.ub.uni-koeln.de/4353/
60. Stutzmann, D.: Système graphique et normes sociales : pour une analyse électronique des écritures médiévales. In: Medieval Autograph Manuscripts. Proceedings of the XVIIth Colloquium of the Comité International de Paléographie Latine, held in Ljubljana, 7–10 September 2010, pp. 429–434. No. 36 in Bibliologia, Brepols, Turnhout (2013). https://www.brepolsonline.net/doi/10.1484/M.BIB.1.101494
61. Stutzmann, D.: Clustering of medieval scripts through computer image analysis: towards an evaluation protocol. Digital Medievalist **10** (2016). https://doi.org/10.16995/dm.61
62. Stutzmann, D.: Ecmen (2017). https://github.com/oriflamms/ECMEN
63. Stutzmann, D.: Les «manuscrits datés», base de données sur l'écriture. In: De Robertis, T., Giovè Marchioli, N. (eds.) Catalogazione, storia della scrittura, storia del libro. I Manoscritti datati d'Italia vent'anni dopo, pp. 155–207. SISMEL - Edizioni del Galluzzo, Firenze (2017)
64. Stutzmann, D., Helias-Baron, M.: ICDAR 2017 competition on the classification of medieval handwritings in Latin script - Dataset (2017). https://zenodo.org/record/5527690
65. Tang, Y., Wu, X.: Text-independent writer identification via CNN features and joint bayesian. In: 2016 15th International Conference on Frontiers in Handwriting Recognition (ICFHR), pp. 566–571. IEEE (2016)
66. Tomiello, A.: Dalla littera antiqua alla littera textualis. Gazette du livre médiéval **29**(1), 1–6 (1996)
67. Wolf, L., Dershowitz, N., Potikha, L., German, T., Shweka, R., Choueka, Y.: Automatic paleographic exploration of genizah manuscripts. In: Fischer, F., Fritze, C., Vogeler, G. (eds.) Kodikologie und Palaographie im Digitalen Zeitalter 2 - Codicology and Palaeography in the Digital Age 2, Schriften des Instituts für Dokumentologie und Editorik, vol. 3, pp. 157–179. BoD, Norderstedt, Germany (2011)
68. Xu, Y., Nagy, G.: Prototype extraction and adaptive OCR. IEEE Trans. Pattern Anal. Mach. Intell. **21**(12), 1280–1296 (1999)

Detecting and Deciphering Damaged Medieval Armenian Inscriptions Using YOLO and Vision Transformers

Chahan Vidal-Gorène[1,2](✉) [iD] and Aliénor Decours-Perez[2]

[1] École nationale des chartes, Université Paris, Sciences & Lettres, Paris, France
chahan.vidal-gorene@chartes.psl.eu
[2] Calfa, Paris, France

Abstract. This paper investigates the development and assessment of a methodology for the automatic detection and interpretation of damaged medieval Armenian inscriptions and graffiti. The research utilizes a newly compiled dataset of 150 images that include a variety of inscriptions, mosaics, and graffiti. These images are sourced from general archaeological site views and vary in quality and type, including drone and archival photos, to replicate real-world database challenges. The results highlight the efficiency of a two-step detection and classification pipeline. The detection phase employs a YOLO v8 model to identify the location and content of inscriptions, achieving an average Precision and Recall of 0.91 and 0.88, respectively. The classification phase uses a Vision Transformer (ViT) to identify similar characters, which outperforms classic CNN-based Siamese networks to handle such a complexity and variation. This approach demonstrates potential for analyzing under-resourced and damaged corpora, thus facilitating the study of deteriorated inscriptions in a variety of contexts.

Keywords: Armenian inscriptions · Digital epigraphy · Computational Paleography · Vision Transformer · Object Detection · Instance Segmentation · Image similarity

1 Introduction

Armenian religious monuments are generally substantially covered with epigraphs and graffiti. They are the writing evidences of the foundation or the renovation of the building, and provide information on the date of construction, the name of the patron, or the purpose of the monument, as well as the passage of the pilgrims. Their deciphering is hindered by the test of time or wilful human damage, which reveals complex even for epigraphists. In some extreme cases, parts or all of the inscriptions are missing and the strong ambiguity of the remaining graphic forms calls for speculation. The most ancient inscriptions that have been dated, prior to the 8th century, though very few, have been the most documented, and can be found in most medieval corpora published since the 19th century, notably in Ališan [2], Yovsēp'ean [24], Kouymjian & Stone [20], Greenwood [8], and Mouraviev [14]. However, the later inscriptions and graffiti, despite partial collection, identification, and transcription efforts since 1966 [4], are often only briefly documented and deciphered due to their sheer volume. The multiplication of image databases, for example within collaborative projects like WikimediaCommons, institutional ones like MonumentWatch, or photogrametric projects carried out by researchers [13] or by private initiatives (Iconem, TUMO), is contributing to the preservation of these testimonies. However, because of their volume, the variety and quality of their formats (written, photographic, 3D), and the damage they have suffered, in the end, very little analysis is done, and the inscriptions remain therefore inaccessible.

This overall inaccessibility is also due to the specific features of the Armenian epigraphic tradition, which encompasses a previous state of the language (or Classical Armenian, even Middle Armenian for some inscriptions), a combination of letters specific to each lapicide – hence, the large variety of monograms –, and the very location of the inscriptions on the monument, which are often high up or poorly oriented (either from the outset or following restoration) and therefore out of reach of the human eye, whether specialist or not (see *infra* Sect. 3). To this day, no computational approach has yet been explored for Armenian epigraphy, therefore, the aim of this paper is first to explore the feasibility of automatic detection of Armenian inscriptions and graffiti in images of varying resolution, and then to explore the classification of the glyphs detected, with a view to proposing an aid for reading these witnesses. This article is an opportunity to build up a first small dataset representative of the Armenian epigraphic production, limited to a capital script from the *erkat'agir* type. This study is set in the context of limited data.

2 Related Works

As a matter of fact, the main difficulty encountered upon computational processing of Armenian sources is the critical lack of data, often incompatible with the training of ML models. If the data creation and retrieval chains for HTR issues regarding medieval handwritten sources are now well established, it is not yet the case for the palaeographical and epigraphical issues, all the more so when sources are damaged. A recent study on Aramaic inscriptions [1] demonstrated that using Generative Adversarial Networks (GANs) to generate 250,000 damaged samples can simulate a representative dataset. This approach achieved over 95% classification accuracy with a common ResNet, given a limited number of different classes (22 Aramaic characters). Although GANs have proven useful, their effectiveness remains primarily within in-domain studies that can tolerate potential overfitting [1,21]. Generally, these datasets do not exceed a few tens of thousands of samples. In our study, we did not use GANs for data improvement, focusing instead on real-world image variations.

Although direct HTR-type approaches have been tried out, with for example the use of Tesseract for the recognition of Tamul inscriptions [7], the state-of-the-art shows today the predominance of a two-step approach: a first step for the character detection and a second for their classification [23]. Predictably, the detection step is performed as an object detection task by faster-RCNN [18] or YOLO [5,18], with an Accuracy and a Recall in average above 90% for Ancient and Byzantine Greek. However, both approaches remain unstable, easily producing numerous false positives when met with an advanced state of deterioration, with strong ambiguity of the residual forms [18]. To resolve this ambiguity when the inscription has a high density, the use of a U-net – trained to segment semantically at pixel level – seems to achieve promising results on bones engraved with Chinese inscriptions [6]. The same applies for a CNN-Siamese approach based once again on a faster-RCNN [18]. The assessment remains qualitative for the time being. In turn, the classification stage remains largely prospective: a semi-supervised classification predominates, based on the use of a ResNet [1,5] or a VGG [10,22] pre-trained on ImageNet, from which the classification layer is removed in order to cluster the extracted features. This method is not only used to classify characters, but also to date inscriptions (with 85.94% Accuracy for Sinhalese incriptions), to identify Latin [11] and Armenian [22] scripts, or categorise hieroglyphic artifacts [9]. The use of CNN-Siamese networks is also used in Ancient Greek [17] and in Aramaic, that displays a 77% efficient classification, just as a VisionTransformer [15]. These approaches, who are based on similarity, exceed a directly supervised classification [18]. Finally, the detection and partial classification of a set of characters can enable statistical completion of the inscription, which has been used for the first time in Greek with 62% Accuracy and in Latin [3,12].

3 Notions of Armenian Epigraphy

The Armenian alphabet, dating back from the 5th century, is composed of 36 letters (24 letters from the Greek alphabet among which are interspersed letters to cover sounds specific to Armenian), with the addition of two new letters in the 12th century to cover two new sounds /f/ and /o/. The Armenian epigraphic production has largely remained confined to the *erkat'agir* type, a capital bicameral script generally associated with the uncial [20], that is classically used in lapidary inscriptions, ancient graffiti and manuscripts up to the 10–11th centuries – date from which its use in manuscripts declines. The alphabet is also used to write numbers.

The Armenian inscriptions are not very difficult to read: the letters are of substantial size, and if there is no spaces in between words and no apparent logic to the word breaks by default, all letters are evenly spaced out at a width equivalent to an upstroke. Writing lines are sometimes even perceptible. The inscriptions can be V-shaped or flat-bottomed engraved (the latter is more represented in inscriptions located up high on the monument). Some churches are heavily covered of inscriptions, sometimes contiguous with no semantic continuity. The Armenian characters however are composed of a small number of structural features, aside from round letters, they are limited to an upstroke, a curved or round stroke and a connecting or cross stroke. Only the equivalents of the apex and the ending stroke can be used to determine the orientation of the upstroke or of the round stroke (voir Fig. 1). Deterioration of any kind to the inscription on either side of the median line results in extreme ambiguity (e.g. addition of artifacts that can be confused with the round or cross strokes, or even the disappearance of a round or cross stroke critical to decipher the letter), that can only be resolved through the understanding of the context. Depending on the lapicide, the serif of the structural strokes can be enhanced with aesthetic features (e.g. triangular engraving).

Fig. 1. Schematic representation of structural features of the main Armenian letters, inspired by the model of the inscription of Mastara (7th c.). The first group is composed of one ascender, one right curved stroke and one right cross stroke; the second group is composed of one upstroke, one left curved stroke and one right cross stroke; the third group is composed of one descender, one right curved stroke and one right cross stroke; the fourth and fifth group are composed of two upstrokes (one ascender and one descender), or possibly of one left upstroke and one right cross stroke; and the sixth group is composed of one upstroke and two curved strokes on each side.

Another difficulty lies in the frequent use of ligatures by the lapicide, who is joining together one or more vowels with one or more consonants for space-saving purposes. The defining structural stroke for an Armenian letter being the upstroke, it can indeed be used as a basis for one or several characters.

The most common ligature shape encloses two letters, but there is no structural limitation to the number of possible combinations. Although, for some combinations, the ligature is easy to read for the epigraphist, for other combinations, it is less self evident and it creates an object detection problem in its own right: should a separate distinct class be created for each ligature? The variety and the low representation of each combination do not support this solution. Conversely, should the annotation strictly follow the 36+2 Armenian letters? The overlapping of bounding boxes (bboxes) and the presence of multiple valid interpretations can cause a high false-positive rate. Moreover, the geometric complexity of ligatures and the independent reading order of the signs add to the challenge (see Fig. 2).

Figure 2 presents two types of common ligatures: the first one consists in the adjunction of the tops and bottoms of the upstrokes with no impact to the reading direction. The strict annotation per character is possible, but requires the blue bbox to have a shorter height, a format likely to be under represented, for which it is reasonable to assume that the model will tend to detect the whole height of the letter, that will become ambiguous. The second example is more

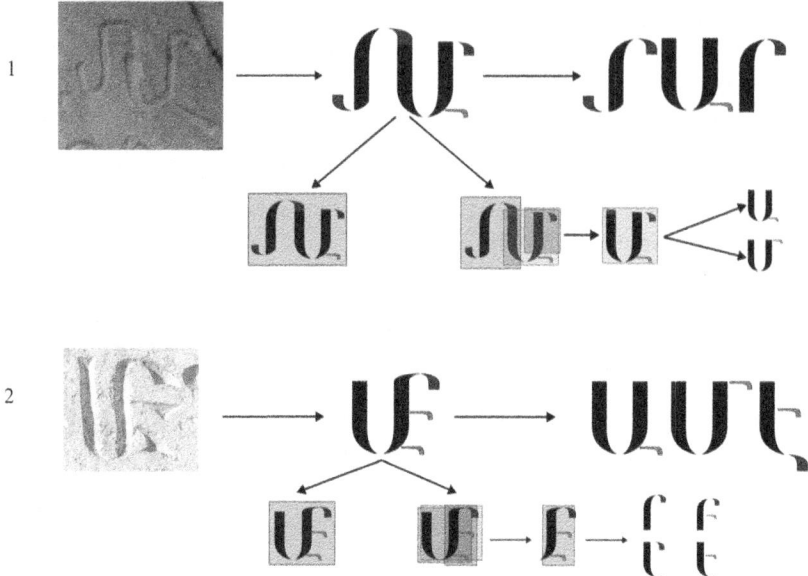

Fig. 2. Two examples of ligatures, their reading, possible annotations and the consequences classification-wise. (1) Inscription of Arudj (8th c.) et (2) inscription of Yeritsmangants (1691). (Color figure online)

frequent: a single upstroke combining all curved and cross strokes. As a result, the green and red bboxes have an intersection close to 1, and the blue bbox may correspond to 4 different shapes, depending on the thickness given by the lapicide to the connecting strokes and serif.

As for the punctuation and the accentuation marks, there is less volume and variety than in Armenian manuscripts. Three types of points are generally found: the median and upper point that indicate a pause, and the two points that marks the ending of the sentence. They come either in diamond or triangular shape, with rectilinear V-shaped engraving. They can be confused with the serif of some characters (orange and red strokes in Fig. 1, when the connecting strokes are fine), or with impacts on the stone.

The medieval graffiti display the same attributes as the inscriptions, with a shallower engraving – the stone being usually scratched –, less regularity in the character morphology, and less rectilinear text. They are often limited to an isolated surname, but can also be part of a dedication such as "Remember me [surname]". Medieval graffiti are scarcely found on the walls of medieval churches (also due to the test of time and reconstructions), on the other hand, walls are covered in very recent graffiti of interest not targeted by the present study.

4 Methodology and Dataset

In order to detect and read the damaged inscriptions and graffiti we are setting a two-step pipeline: first, detection of the insription and of the characters; then, a classification of the crops of caractères obtained via similarity calculation.

4.1 Dataset

The dataset is composed of 150 images of inscriptions, mosaics and various graffiti. The stress is put on the inscriptions prior to the 9th century and deteriorated, hard to read, as well as Armenian graffiti from Sinai from the 5–7th centuries [16,19]. Around a quarter of the inscriptions are dated from after the 9th century, with an upper limit in the 17th century, in order to have more legible images. The images used are not all centered on the inscriptions but are general views of walls or overall views of an archaeological site, and mix recent views with archive photos. The goal of this variety is to simulate a real-life application, with a wide range of shots and qualities (brightness, zoom, sharpness, camera, etc.), representative of existing digital databases (see Fig. 3). For example, it is the case of images 3 and 5 in Fig. 3, shot in the 1970s, or conversely, image 6 is a 4K view taken by drone to create a photogrammetric model. This nevertheless represents a very significant bias in the ability of the models to converge, especially given the volume considered in the article.

Figure 3 also underlines the difficulty encountered by the dataset with regard to the ambiguity that exists between a deterioration mark and a character (e.g. images 2 and 7), and among graffiti, when mixed with other non-Armenian graffiti (e.g. image 5, where Greek and Latin graffiti are not tagged, like in image 6 if non-Armenian inscription or unrelevant graffiti are present in the overall view).

Table 1 summarizes the distribution of images according to the source type. Mosaics, fairly scarce in the Armenian production, are largely minority within the dataset. There are perhaps as many inscriptions as graffiti, but the count overall is clearly in favor of lapidary inscriptions, the latter generally have more than 3 lines each and more text.

Table 1. Dataset summary

Source	Type	Date	Damage	Im. Quality	Script	Insc.	Lines	Char.
Inscription	Mixed (drone, camera, book)	7–17th c.	Partial	Mixed	Erkat.	92	303	6,072
Mosaïc	Camera	5–6th c.	No	NTR	Erkat.	9	10	352
Graffiti	Archives	5–6th c.	Yes	Blur, overexposure	Erkat.	74	145	870
TOTAL: 150 images including						175	458	7,294

Fig. 3. Dataset samples. (1) Door inscription from Aruj monastery (7th c.), (2) Komitas inscription (618 A.D.) from Album of Armenian Paleography, (3 and 5) graffiti of Sinai (5th c.) from The Rock Inscriptions Project, (4) Inscription of Grigoras from Mastara (7th c.), (6) West front of Mastara (7th c.), (7) Yakovb inscription of Ereruyk (6–7th c.) and (8) Uxtatur inscription from Talin (683 A.D.)

At the annotation level, we decided to resolve the ambiguity of ligatures mentioned in Sect. 3 by identifying a type ligature as a separate class (each different ligature is now the subject of a separate class). Predictably, the dataset is very unbalanced in terms of character classes: the letter a, the most frequent in the alphabet, covers 26% of annotation on its own, whereas the ligatures and the characters $\bar{o}$ and f have only a single occurrence, and the characters $č$, $ž$, and $š$ have fewer than 10 occurrences.

4.2 Step 1: Three-Stage YOLO-Based Inscription Detection

At the detection level, we are evaluating the use of YOLO v8, through the combination of three models:

1. m-Inst$_{insc}$: detection of inscriptions in an image (instance segmentation, for management of various viewing angles and separation from adjacent inscriptions). The output of this model will correspond to the input of the following models;
2. m-Obj$_{char}$: detection of glyphs through an object detection approach with a single class (there is not enough representatives per class for direct detection);
3. m-Inst$_{line}$: detection of a line of text to re-assemble and order glyphs (instance segmentation). Line characters are sorted using the centroid, with simple left-to-right applied to characters, up-and-down applied to lines (Fig. 4).

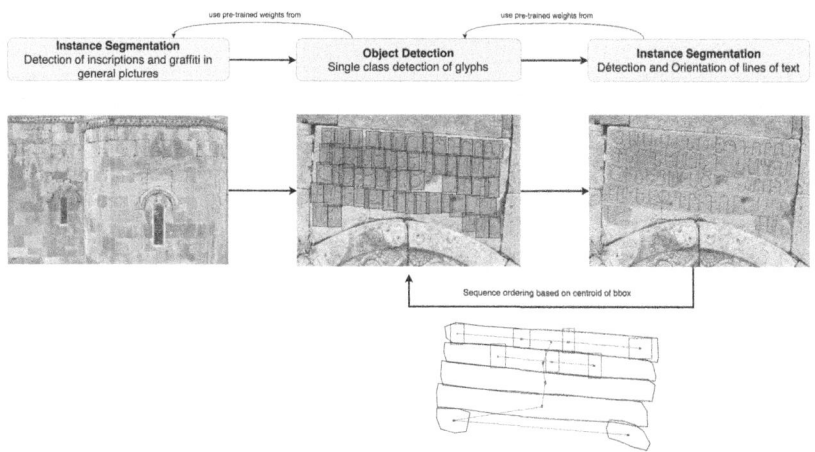

Fig. 4. Detection pipeline involving three YOLO v8 models

All three models perform a single-class classification. For all three, intense dynamic data augmentation is performed during training, similar to that used for papyri detection [18], including image scaling at each iteration and random pixel dropout to simulate further deterioration. The dropout is set at 20% and the size of image at 1024px. We use the weights from the previous model as the initial weights for the next model, effectively fine-tuning each subsequent model based on the trained weights of its predecessor. This approach aims to incrementally improve performance by leveraging the learning from earlier stages. At each step, 10 cross-validation are performed by redistributing the data into train and val, and then evaluation occurs on the same fixed test set. The scores presented in Table 2 are the average obtained of these 10 experiences.

4.3 Step 2: Classification for Characters Similarity

At the classification level, we are following the approaches used for the Aramaic bowl [15], though changing the task performed and the depth of the models.

The first experiment consists in training a Siamese network to identify pairs of similar images (see Fig. 5). The architecture of our Siamese network consists of three main components: a pretrained ResNet50 that will encode an embedding, an euclidian distance for evaluating the similarity, and the contrastive loss function that penalizes the model if the distance between similar inputs exceeds 0.5 or if it falls below this threshold for dissimilar inputs, thus promoting the generation of appropriate embeddings. One of the models of the Siamese network is trained with similar pairs and the other with dissimilar pairs. The contrastive loss function computes the loss for each pair of embeddings, encouraging similar pairs to have a smaller distance and dissimilar pairs to have a larger distance. The general formula for contrastive loss is as follows: $L = (1-y) \times Dist_{Eucl}^2 + y \times max(0, \sigma - Dist_{Eucl})^2$, with $y \in \{0,1\}$, 0 for similar and 1 for dissimilar, and σ the threshold for dissimilarity.

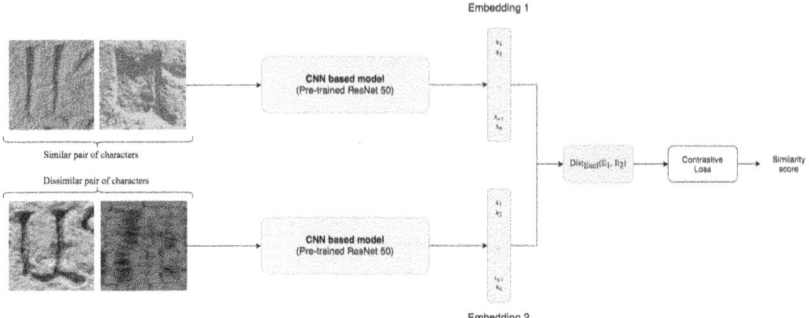

Fig. 5. CNN-based Siamese network using a contrastive loss for similarity classification of characters

The second experiment relies on the use of a distilled Vision Transformer (Data-efficient Image Transformers or DeiT). The approach is no longer based on CNNs but on a full transformer based approach. Images are divided into fixed-size, non-overlapping patches to input into the Transformer encoder. These patches are linearly embedded, and a class token is added as a global image representation for classification. Additionally, absolute position embeddings are incorporated, and the assembled vector sequence is processed by a standard Transformer encoder. Here, the Transformer model's attention mechanism is designed to capture global dependencies and modeling long-range interactions between image patches. The aim is not to train the model with similar and dissimilar pairs anymore, but to create an index from the embeddings of all datasets, and then to perform a similarity search for a given input (see Fig. 6).

4.4 Metrics

To measure the effectiveness of the entire pipeline for detection and reading, we define a comprehensive metric that combines the individual evaluation metrics (Precision, Recall, and mean Average Precision, mAP) from each step into a single score. The F1-Score ($F1_i$) at each stage balances Precision and Recall, providing a robust measure that accounts for both true positive detection and the minimization of missed relevant data. The mean Average Precision (mAP_i) evaluates the model's precision at different Intersection over Union (IOU) thresholds, reflecting its consistency and accuracy.

Weighting factors (w_i) are assigned to each stage to reflect their relative importance in the pipeline. For instance, we prioritize the detection of inscriptions and characters over lines. We combine these metrics with a weighted sum, balancing the F1-scores and mAPs according to their relevance in the model type. This weight is noted α_i. The final score, Global Score (GS), is the sum of each weighted score (S_i). Thus, for $i \in [1, n]$:

$$GS = \sum_{i=1}^{n} w_i \times S_i = \sum_{i=1}^{n} w_i \times (\alpha_i \times F1_i + (1 - \alpha_i) \times mAP_i) \quad (1)$$

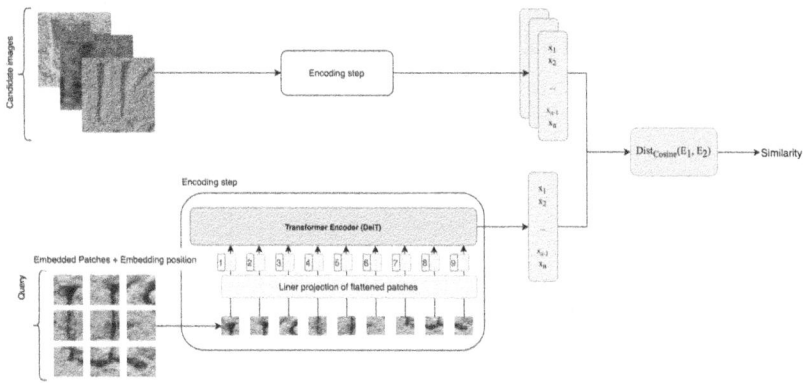

Fig. 6. Vision Transformer using embedded patches and DeiT encoder to perform similarity

5 Results and Discussion

Table 2 summarizes the outcomes achieved for each detection step. Several configurations have been tested: (i) with or without data augmentation, (ii) single and multi-class for the character detection, and (iii) with or without the use of the output obtained after step 1 (inscription detection).

We observe that data augmentation has a highly variable impact on improving results. In general, it is worth mentioning that it increases the Precision, with a limited effect on the Recall (thus producing more false positives), except for the single-class configuration for character detection (standalone), where data augmentation results in decreased both Precision and Recall (from 0.87 to 0.73 in Precision and from 0.54 to 0.46 in Recall), which could suggest that the augmentation method used may not be optimal or the model overfits without it.

At the character and line detection level, using the output obtained after step 1 (inscription detection) results in a largely better detection, around +30%, even if it is not yet optimal, and the mask produced tends to cut the edges of the inscriptions, thus cropping some of the letters.

Regarding the character detection, we get mixed results in multi-class detection, to be put in perspective with the classification task. Nevertheless, the model m-Obj$_{char}$ multi-class 4 (using m-Inst$_{insc}$ output) achieves 0.43 Precision and 0.47 Recall in detection and direct classification of characters, on the same basis as the results obtained on the papyri with the very same model [18], but with five time less data in training (35,597 characters for the papyri vs 7,294). However, the mAP is low. The scores obtained for YOLO multi-class are the average of all classes and therefore reflect only the few most endowed classes (see *infra* 4.1).

As for the results for similar pair identification, in order to overcome the extreme disproportion between classes, we keep only 50 samples per class. The under-resourced classes are artificially augmented through data augmentation (using blur, rotation, channel dropout, and pixel dropout), and each experiment is repeated 10 times with random sample selection for each class. Table 3 summarizes the results.

Table 2. Inscription detection results using several data augmentation and multi/single-class configuration. Mean Precision, Recall and mAP after 10 random splits

	Augment	P	R	mAP
Inscription detection (mask)				
m-Inst$_{insc}$ 1	-	0.72	0.87	0.73
m-Inst$_{insc}$ 2	✓	**0.91**	**0.88**	**0.90**
Char detection (bbox)				
m-Obj$_{char}$ multi-class 1 (stand alone)	-	0.21	0.18	0.16
m-Obj$_{char}$ multi-class 2 (stand alone)	✓	0.29	0.17	0.24
m-Obj$_{char}$ multi-class 3 (using m-Inst$_{insc}$ output)	-	0.43	0.47	0.28
m-Obj$_{char}$ multi-class 4 (using m-Inst$_{insc}$ output)	✓	0.57	0.61	0.54
m-Obj$_{char}$ single-class 1 (stand alone)	-	0.87	0.54	0.77
m-Obj$_{char}$ single-class 2 (stand alone)	✓	0.73	0.46	0.69
m-Obj$_{char}$ single-class 3 (using m-Inst$_{insc}$ output)	-	0.90	0.82	0.89
m-Obj$_{char}$ single-class 4 (using m-Inst$_{insc}$ output)	✓	**0.91**	**0.84**	**0.90**
Line detection (mask)				
m-Inst$_{line}$ single-class 1 (stand alone)	-	0.67	0.34	0.64
m-Inst$_{line}$ single-class 2 (stand alone)	✓	0.74	0.31	0.65
m-Inst$_{line}$ single-class 3 (using m-Inst$_{insc}$ output)	-	0.94	0.91	0.95
m-Inst$_{line}$ single-class 4 (using m-Inst$_{insc}$ output)	✓	**0.94**	**0.93**	**0.96**

Table 3. Classification results. Accuracy corresponds to the performance of the model in training with manually labeled data, where Precision and Recall correspond to the test using crops of the detection step

Model	P	R	Acc$_{similar}$	Acc$_{dissimilar}$
Siamese	0.42	0.42	0.54	0.55
ViT	0.67	0.71	0.78	0.81

The results of the Siamese network in this configuration are much lower than those of the Vision Transformer, and even equivalent to those of YOLO in multi-class detection. The limits of the Siamese network can undoubtedly be explained by the very small size of the dataset due to its complexity and it is appropriate, at this stage, not to completely exclude it from the experiments. The ViT, on the other hand, shows an excellent ability to identify similar pairs in under-resourced contexts, including when the damage or support vary a lot. Table 4 gives the global score for each pipeline.

Table 4. Global scores for each configuration of Detection and Classification, with the sum of $w_{detection,i} = 0.6$ and $\alpha = 0.3$ to prioritize mAP in detection tasks

Category	w	α	P	R	F1	mAP	Score
Inscription detection	0.3	0.3	0.91	0.88	0.89	0.90	0.90
Char. Detection single-class	0.25	0.3	0.91	0.84	0.87	0.90	0.89
Char. Detection multi-class	0.45	0.3	0.43	0.47	0.45	0.28	0.33
Line detection	0.05	0.7	0.94	0.93	0.93	0.96	0.94
Siamese	0.2	1.0	0.42	0.42	0.42	0.00	0.42
ViT	0.2	1.0	0.67	0.71	0.69	0.00	0.69
Combined Results							
Ins+CharSingleClass+Line+Siamese							0.62
Ins+CharSingleClass+Line+ViT							**0.68**
Ins+CharMultiClass+Line							0.65

The final results show a slight advance of the YOLO + VisionTransformer configuration for the detection and identification of characters in damaged Armenian inscriptions and graffiti. Direct detection by YOLO also seems to be possible, but the disproportion of classes between YOLO multi-class and ViT skews the comparison.

6 Conclusion

The article proposes and assesses an end-to-end pipeline for detecting and reading Armenian inscriptions in a very under-resourced and damaged context. The detection scores demonstrate the benefit of a three-step detection of inscriptions, characters and lines, with an average Precision of 0.92 and an average Recall of 0.88. The task of character classification, considered as a task of identification of similar/dissimilar images, due to the high ambiguity of the characters and the small size of the dataset, remains prospective but the use of a Vision Transformer outperforms a classic CNN-Siamese network. The Accuracy obtained on the test set is on average 0.79, with a Precision of 0.67 and a Recall of 0.71 in real conditions. For the future, we plan to significantly strengthen the dataset in order to increase the representativeness of under-resourced classes and to increase the versatility of the model in real images. Incorporating a language model could further improve the results by providing contextual understanding of the inscriptions. For instance, using a pretrained Armenian language model might help disambiguate characters based on surrounding text. This approach is worth exploring in future work to enhance the accuracy and reliability of our detection and classification pipeline.

Disclosure of Interests. The authors have no competing interests to declare that are relevant to the content of this article.

A Appendix

See Fig. 7.

Fig. 7. Qualitative results of inscription detection, single-class character detection and character similarity using ViT

References

1. Aioanei, A.C., Hunziker-Rodewald, R.R., Klein, K.M., Michels, D.L.: Deep Aramaic: towards a synthetic data paradigm enabling machine learning in epigraphy. PLoS ONE **19**(4), e0299297 (2024)
2. Ališan, L.: Ayrarat. S. Łazar, Venise (1890)
3. Assael, Y., et al.: Restoring and attributing ancient texts using deep neural networks. Nature **603**(7900), 280–283 (2022)
4. Collective: Diwan hay vimagrut'iwn (= Corpus of Armenian inscriptions). National Academy of Sciences of Armenia, Erevan (1966–2017)
5. Eyharabide, V., et al.: Study of historical byzantine seal images: the bhai project for computer-based sigillography. In: Proceedings of the 7th International Workshop on Historical Document Imaging and Processing. HIP '23, pp. 49–54. Association for Computing Machinery, New York, NY, USA (2023). https://doi.org/10.1145/3604951.3605523
6. Fu, X., Zhou, R., Yang, X., Li, C.: Detecting oracle bone inscriptions via pseudo-category labels. Herit. Sci. **12**(1), 107 (2024). https://doi.org/10.1186/s40494-024-01221-5
7. Giridhar, L., Dharani, A., Guruviah, V.: A novel approach to OCR using image recognition based classification for ancient Tamil inscriptions in temples. arXiv preprint arXiv:1907.04917 (2019)
8. Greenwood, T.W.: A corpus of early medieval Armenian inscriptions. Dumbarton Oaks Papers **58**, 27–91 (2004)
9. Hayon, O., Münger, S., Shimshoni, I., Tal, A.: Arcaid: analysis of archaeological artifacts using drawings. In: Proceedings of the IEEE/CVF Winter Conference on Applications of Computer Vision, pp. 7264–7274 (2024)
10. Heenkenda, H., Fernando, T.: Chronological attribution of Sinhalese inscriptions using deep learning approaches. J. Natl. Sci. Found. (2023). https://doi.org/10.4038/jnsfsr.v51i3.11200
11. Kestemont, M., Christlein, V., Stutzmann, D.: Artificial paleography: computational approaches to identifying script types in medieval manuscripts. Speculum **92**(S1), S86–S109 (2017)
12. Locaputo, A., Portelli, B., Colombi, E., Serra, G., et al.: Filling the lacunae in ancient latin inscriptions. In: IRCDL, pp. 68–76 (2023)
13. Magarditchian, A., Vidal-Gorène, C.: L'apport de la photogrammétrie à des prospections archéologiques et paléographiques en Arménie. Études arméniennes contemporaines **14**, 163–183 (2022)
14. Mouraviev, S.: Erkataguir: ou Comment naquit l'alphabet arménien. Academia Verlag, Sankt Augustin (2010)
15. Naamneh, S., et al.: Classifying the scripts of Aramaic incantation bowls. In: Proceedings of the 7th International Workshop on Historical Document Imaging and Processing. HIP '23, pp. 55–60. Association for Computing Machinery, New York, NY, USA (2023). https://doi.org/10.1145/3604951.3605510
16. Negev, A.: The inscriptions of wadi Haggag, Sinai. Qedem **6**, 1–100 (1977)
17. Sajjad, H., Siddiqi, I., Moetesum, M., Marthot-Santaniello, I.: Learning structural similarities from handwriting on papyri - an application to scribe characterization. In: 2023 International Conference on Frontiers of Information Technology (FIT), pp. 31–36 (2023). https://doi.org/10.1109/FIT60620.2023.00016

18. Seuret, M., Marthot-Santaniello, I., White, S.A., Serbaeva Saraogi, O., Agolli, S., Carrière, G., Rodriguez-Salas, D., Christlein, V.: ICDAR 2023 competition on detection and recognition of Greek letters on papyri. In: Fink, G.A., Jain, R., Kise, K., Zanibbi, R. (eds.) ICDAR 2023. LNCS, vol. 14188, pp. 498–507. Springer, Cham (2023). https://doi.org/10.1007/978-3-031-41679-8_29
19. Stone, M.: Armenian Inscriptions from Sinai: Intermediate Report with Notes on Georgian and Nabatean Inscriptions. Harvard Armenian Texts and Studies. Maitland Publications (1979)
20. Stone, M.E., Kouymjian, D., Lehmann, H.J.: Album of Armenian Paleography. Aarhus University Press, Aarhus (2002)
21. Vidal-Gorène, C., Camps, J.B., Clérice, T.: Synthetic lines from historical manuscripts: an experiment using GAN and style transfer. In: Foresti, G.L., Fusiello, A., Hancock, E. (eds.) ICIAP 2023. LNCS, vol. 14366, pp. 477–488. Springer, Cham (2023). https://doi.org/10.1007/978-3-031-51026-7_40
22. Vidal-Gorène, C., Decours-Perez, A.: A computational approach of armenian paleography. In: Barney Smith, E.H., Pal, U. (eds.) ICDAR 2021. LNCS, vol. 12917, pp. 295–305. Springer, Cham (2021). https://doi.org/10.1007/978-3-030-86159-9_20
23. Vijayalakshmi, R., Gnanasekar, J.: A review on character recognition and information retrieval from ancient inscriptions. In: 2022 8th International Conference on Smart Structures and Systems (ICSSS), pp. 1–7 (2022). https://doi.org/10.1109/ICSSS54381.2022.9782241
24. Yovsēp'ean, G.: K'artēz hay hnagrut'ean (= Armenian Paleography Atlas). Šołakat' **1**, 170–214 (1913)

Optimizing HTR and Reading Order Strategies for Chinese Imperial Editions with Few-Shot Learning

Marie Bizais-Lillig[1,2], Chahan Vidal-Gorène[3,4(✉)], and Boris Dupin[3]

[1] UR-1340 GÉO and USIAS, Université de Strasbourg, Strasbourg, France
[2] Huma-Num Consortium DISTAM, CNRS, Strasbourg, France
[3] Calfa, Paris, France
chahan.vidal-gorene@chartes.psl.eu
[4] École nationale des chartes, Université Paris, Sciences and Lettres, Paris, France
https://www.calfa.fr/

Abstract. In this study, we tackle key challenges in layout analysis, reading order, and text recognition of historical Chinese texts. As part of the CHI-KNOW-PO Corpus project, which aims to digitize and publish an online edition of 60,000 xylographed documents, we have developed and released a specialized small dataset to address this common issues in HTR of historical documents in Chinese. Our approach combines a CNN-based instance segmentation model with a local algorithmic model for reading order, achieving a mean precision of 95.0% and a recall of 93.0% in region detection, and a 97.81% accuracy in reading order. Text recognition is conducted using a CRNN model enhanced with GAN-augmented data, effectively addressing few-shot learning challenges with an average accuracy of 98.45%, demonstrating the effectiveness of a small and targeted dataset over a large-scale approach. This research not only advances the digitization and analytical processing of Chinese historical documents but also sets a new benchmark for subsequent digital humanities efforts.

Keywords: Chinese · HTR · Historical Documents · Layout Analysis · Reading Order · Dataset

1 Introduction

Most experiments in the field of text recognition applied to Chinese language focus on modern and contemporary texts onwards (beginning at the very end of the 19th century). Characters are not so diverse, texts are punctuated, and the layout generally differs, with many texts written on horizontal lines from left to write.¹ There have been experiments to try and extract text from ancient editions similar to the ones used in the present dataset. However, the datasets produced during the projects often stay private when they are not published under license. In Taipei, the Academia Sinica started developing models to extract text from images since the early 2000 s. The rich collection² of texts neatly edited by this institution, which many sinologists use on an every-day basis, is, however, mostly accessible by subscribed membership, and its content is licensed. The Chinese Text database is another commonly used resource, freely accessible online, but whose content is also submitted to copyright limitations. Also, despite the progress made in the field of text recognition³, available models are often so specialized that they don't seem very useful in other contexts, and training datasets are not so common.⁴ In other words, the field of imperial Chinese studies is in need of an open dataset of texts and images, produced with the help of HTR technologies.

Within the scope of the CHI-KNOW-PO Corpus project, which aims to establish a comprehensive online and searchable corpus of 60,000 pages, we are experimenting with the development of models to address specific challenges associated with late imperial China editions. The project's goals include the definition of an efficient pipeline, the publication of a training dataset, and the availability a broader corpus of texts from the Chinese first millennium. The key challenges include: 1/ precisely analyzing and structuring page layouts to accurately order the full text, and 2/ recognizing a vast array of different glyphs. Despite years of research in this domain across both Eastern and Western contexts-as noted by recent studies from Fudan University (Shanghai) [31]-automatic extraction of ordered text based on images of Imperial China xylographed editions remains highly experimental and in need of established benchmarks like those the CHI-KNOW-PO Corpus project aims to provide.⁵ To tackle these challenges effectively, we have created a specialized small dataset tailored to the project's needs, that can be used as a benchmark dataset for HTR purposes Chinese historical documents. We propose some results as a baseline for this dataset.

---

[1] Layout may still be complex, when working on newspapers for instance. For this type of data, see the Heidelberg Centre for Transcultural Studies project [11] and dataset [10].

[2] The database is called Scripta sinica. Url: https://hanchi.ihp.sinica.edutw/ihp/hanji.htm (Last accessed on May 5th 2024).

[3] See for instance the ICDAR 2017 Competition on Reading Chinese Text in the Wild [21] along with Donald Sturgeon's recent publication [22].

[4] Certainly, the Kanripo repository(https://www.kanripo.org/ (last accessed May 5th 2024).), which represents a huge amount of text mirroring black and which images of the *Siku quanshu* 四庫全書 is very interesting. Nevertheless, the quality of the images reduces the interest of such dataset.

[5] The Fudan team contributes to such benchmarks; see https://github.com/FudanVI/benchmarking-chinese-text-recognition (last accessed May 5th 2024). Unlike other initiatives, however, the CHI-KNOW-PO Corpus project adheres to the FAIR principles.

2 Sources: Corpus Characteristics and Dataset

The corpus was circumscribed according to several criteria, the main objective being to represent a literate library of the first millennium-excluding Buddhist texts, which are collected within the framework of the collaborative project on Medieval China Buddhist texts directed by Christoph Anderl[6]. Exhaustivity once Buddhist texts had been put aside was still an impossible ideal. Most Classics-Confucian or Taoist in particular-commented on by scholars of the first millennium as well as historical works already available online have hence been excluded. We favored thematic coherence: our research project explores the phenomena of co-occurrences and repetitions or echoes between passages of texts which refer to plants. This prism allows us to include a wide variety of genres: knowledge texts such as lexicons and compendiums classified by categories (*leishu* 類書 in Chinese, often referred to under the misleading term encyclopedia), poetry, treatises on *materia medica*, and agricultural treatises in particular. The variety of genres is accompanied by a relative variety in form. Certain works are extracted from large collections compiled either on imperial order or privately by a bibliophile, while others constitute independent physical units. Texts also vary in form, length, and organization: the corpus includes lexicons, anthologies of poems, along with chaptered prose texts. Finally, some texts are commented on by one or more exegetes, while others are the work of a single author. Before illuminating more precisely the issues that the glosses represent in this project, let us describe the dataset and the material and formal characteristics of the corpus as a whole.

2.1 Qualitative Description of Targeted Documents

The CHI-KNOW-PO Corpus project contains three types of texts, from which we extracted a sample in order to construct a dataset for model training adapted to its processing.

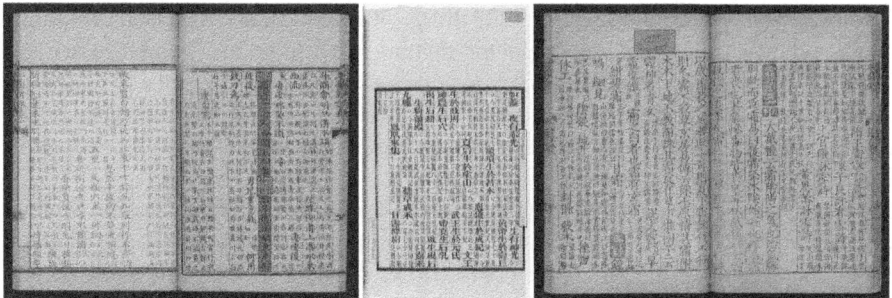

Fig. 1. Overview of the dataset, with typical page layout of a xylographed edition on the left and on the right. Main semantics zones and lines are displayed.

---

[6] https://www.database-of-medieval-chinese-texts.be/ (last accessed March 13 2024).

Anthologies. The *Li Shan zhu Wenxuan* 李善注文選 (The Selection of Belletristic texts commented upon by Li Shan, abbreviated to Li Wenxuan, n°A-1) is a large anthology of 30 *juan* (rolls), organized by genre, and compiled by Xiao Tong 蕭統 (501–531), Prince Zhaoming 昭明太子, under the Southern dynasty of the Liang 梁 (502–557). It comprises more than 700 texts composed in many different genres between Antiquity and the early Medieval period. Each of these texts stood as a model for later compositions by literati. More than half of the anthology is made of *shi* 詩 poetry and *fu* 賦 rhymed prose, which were refined texts and whose vocabulary was very rich. To shed light on the meaning of the texts and their intertextual relations with other reference texts, these texts were commented upon less than a century after its compilation. The earliest commentary that survived was Li Shan's 李善 (630–689), which also became the most famous. Fragments of the early versions of the text can be found in the Dunhuang collections of the Bibliothèque nationale de France. Complete versions transmitted to us are relatively late. This one was printed in 1809 based on xylographic plates from the Song dynasty (960–1279). The work is composed of four cases of six leaflets each. It is 30 cm high and preserved at the National University Library in Strasbourg (BNU, call number: FP.12.4.0001)[7].

The *Liuchen zhu Wenxuan* 六臣注文選 (abbreviated Liuchen Wenxuan, n°A-2) corresponds to the same book. After Li Shan commented upon the *Wenxuan*, a group of scholars considered his commentary too elitist and difficult to understand. The collection that merges the commentaries produced by the five opponents to Li Shan with the ones by Li Shan is called the *Selection of Belletristic texts commented upon by the six ministers*. This text is interesting because of its pedagogical input and its richness. The edition used in the dataset is a rather late edition, printed in 1923. It belongs to the *Sibu congkan* reference collection. The work is composed of three cases of ten leaflets each. Its is 20 cm high and is preserved at the BNU (FP.12.8.0001). It contains manual annotations in the margins.

The *Yutai xinyong* 玉臺新詠 (New Songs of the Jade Terrace, abbreviated Yutai, n°A-3) is a collection of 769 court poems from Antiquity and the Early Medieval period compiled by Xu Ling 徐陵 (507–583). The collection is famous for its thematic coherence, most poems dealing with love and women. This is a rather late edition, printed in 1879, precious because of the presence of commentaries. It consists of one case of ten 18 cm high leaflets. It is preserved at the Library of the Institut des hautes études chinoises at the Collège de France in Paris (BIHEC, call number: V XIV 69˙ (1–8)).

The *Quan Tang shi* 全唐詩 (Complete Poems of the Tang dynasty, abbreviated Tangshi, n°A-4) corresponds to the complete collection of Tang dynasty poems compiled during the Qing (1644–1911) dynasty. It contains some 50,000 poems. For each author, biographical elements are provided in the form of a commentary. This one is one of the earliest editions (possibly the first), printed in 1707. This work is composed of twelve boxes of ten leaflets each. It is 16.5 cm high and is preserved at the BIHEC (SB 4002 (1–12) (1–120)).

---

[7] The Libraries part of the CHI-KNOW-PO Corpus project are listed in the acknowledgement section.

Scholarship. The *Beitang shuchao* 北堂書鈔 (Written excerpts from the Northern Hall, abbreviated Beitang, S-1) is one of the earliest *leishu* composed in China. It is divided in categories, while each category is composed of a series of entries. Each entry includes 1/ a series of definitions (usually excerpted from ancient reference books), and 2/ a series of famous quotations (that can be reused by literati in their own writings). This edition, which is a reconstruction of the book, was printed in 1888. The structure of the book is not easily readable. This work is composed of four cases of fives leaflets each. It is 31 cm high and is preserved at the University Library des langues et des civilisations in Paris (BULAC, call number: BIULO CHI.1087).

The *Bowu zhi* 博物志 (Notes on things at large, abrreviated Bowu zhi, S-2) is a short book composed by Zhang Hua 張華 (232–300) during the Western Jin 西晉 (265–316). It is composed of ten chapters. Each chapters stores a series of tales on the world. The book, printed in 1875, consists in one leaflet. It is preserved at the BULAC (BIULO CHI.1140).

The *Chuxue ji* 初學記 (Records to begin studies, abbreviated Chuxue, S-3) is one of the most prestigious *leishu* of Chinese history. It was composed by a team on imperial order during the Tang 唐 dynasty (618–907). It defines categories to classify entries. Each entry is followed by different sections: definitions, citations from texts in different genres, so called "parallel matters". This 1587 edition is very clearly presented. It contains two cases of twelve leaflets each. It is 27.5 cm high and is a preserved at the BIHEC (SB 3701 (1–2)).

The *Erya yintu* 影宋鈔繪圖爾雅 (Illustrated Elegantiae, facsimile of a Song edition, abbreviated Erya, S-4). The *Erya* is a Confucian classic. It is a lexicon, ordered by category. Each entry is followed by a series of equivalents (not exactly synonyms). It is often considered to be essential to decipher the meaning of rare terms used in the Confucian Canon. This edition is characterized by the presence of phonological annotations and by its illustrations. The work was printed in 1801. It consists of three illustrated volumes. Each volume is 34 cm high. It is preserved at the BULAC (BIULO CHI.1938(1)-(3)).

The *Mao Shi caomu niaoshou chongyu shu* 毛詩草木鳥獸蟲魚疏 (Commentary on flora and fauna in the *Poems* according to Mao, abbreviated Maoshi shu, S-5) is a commentary established by Lu Ji 陸璣, who lived between 200 and 500. It is a commentary of the Confucian classic anthology of 300 poems known as the *Shijing* 詩經. The commentary doesn't include its base text. It is composed of entries that are named using one line of one poem. The selected lines all include elements of fauna or flora. This element is defined using multiple sources: it is a compilation of all information available on these elements (where they grow or live, what are their physical characteristics, how they are called in different places and times). This work is part of a larger collection, printed in 1857, and preserved at the BIHEC (V I 111 (1) 5).

The *Yiwen leiju* 藝文類聚 (Compilation of texts and works ordered by category, abbreviated Yiwen, S-6) is, with the *Chuxue ji*, S-3, the other major *leishu* of early Chinese history. It was also composed during the Tang dynasty (618–907), by a team headed by the well-known scholar Ouyang Xun 歐陽詢 (557–

641). This work was printed in 1879. It consists of eight boxes of five leaflets each. It is 20.5 cm high and is preserved at the BIHEC (CIII 5–7 (1–8)).

The *Zhi bu zu zhai cong shu* 知不足齋叢書 (Collection of the One who did not know enough, abbreviated Zhibuzu, S-7) corresponds to what is called *congshu* 叢書 (collection) in Chinese. It means that it is a compilation of a number of books-in this case, 207 books. The *Zhibuzu* collection was compiled by Bao Tingbo 鮑廷博 (1728–1814). It includes all sorts of books, mainly encyclopedias and other books of scholarship, from the first millennium. It is a very large and diverse collection, although our selection means to establish links between the books included in the *Zhibuzu* and other books present in the dataset (poems, texts on plants). This book was printed in 1822. The layout changes from one title to the other. It is composed of 120 leaflets, 30 cm high, and is preserved at the BIHEC (F X 2 (1–15) 1–120)).

Technical and Practical Knowledge. The *Gujin shiwen leiju* 古今事文類聚 (Compilation of texts and activities from present and past ordered by categories, abbreviated shiwen leiju, T-1). This *leishu* is composed of a series of *leishu* from different periods (from the 12th century onwards). Although such a *leishu* resembles scholarship encyclopedias, this one focuses more clearly on practical issues. The work was printed in 1604. It was later bound in a Western way. It is composed of 26 volumes, 27.5 cm high. It is preserved at the BIHEC (SB 3705 (1–26)). It has suffered from humidity.

The *Qimin yaoshu* 齊民要術 (Essential techniques of the people of Qi, abbreviated Qimin yaoshu, T-2) is the earliest treatise on agriculture preserved in Chinese history. It is a short texte. This edition was printed in 1896. It is preserved at the BIHEC (V I 22 (1–4)).

The *Xinzhai shizhong* 心齋十種 (Tens types for the temperate heart, abbreviated Xinzhai, T-3) is a short collection of practical texts including treatises and encyclopedias. It was printed between 1785 and 1788. It consists of one box of four leaflets. It is 16.8 cm high and is preserved at the BIHEC (V I 53 (1)).

Although the selected medieval corpus (ca. 250–1000) covers relatively varied genres, it is generally studied based on imperial editions from the second millennium,[8] and is hence rather homogeneous in its physical form.

We are mostly dealing with xylographed accordion-bound books. This means that the medium is generally thin paper, which is written on only one side, and which is folded to form the equivalent of a sheet in a codex. The side 'trapped' in the fold remains inaccessible to the reader, its function being to reduce the noise created by the back of the sheet when reading the front, and vice versa.

Each page is composed of a regular number of columns, delineated with vertical and horizontal lines-in imitation of the string of bamboo strips on which texts were inscribed in Antiquity. The double-pages, which correspond to a leaf or a sheet, in a xylographed edition are printed from a single plate, and separated by a column of a particular shape (diamond or double fish, in salmon pink in Fig. 1) where metadata are inscribed. This area, which corresponds to the fold of

---

[8] Most manuscripts disappeared in the transmission process.

Table 1. Dataset Distribution

Title	Pages	Lines	Characters
Li Wenxuan, A-1	27	771	16,697
Liuchen Wenxuan, A-2	29	889	16,271
Yutai, A-3	10	590	10,220
Tangshi, A-4	10	493	7,685
Beitang, S-1	35	1,508	27,388
Bowu zhi, S-2	23	302	7,874
Chuxue, S-3	20	1,267	24,411
Erya, S-4	40	2,577	20,121
Maoshi shu, S-5	10	397	8,357
Yiwen, S-6	11	357	7,409
Zhibuzu, S-7	49	1,804	29,524
shiwen leiju, T-1	20	1,055	15,629
Qimin yaoshu, T-2	20	885	18,307
Xinzhai T-3	23	864	14,143
TOTAL	327	13,759	224,036

the sheet, becomes difficult to read in a bound book: we only see half-characters which can correspond to the title of the work, the title of the part, and the number of the sheet (which sometimes starts again at the beginning of each chapter). In the project, despite the poor readability of this area, we decided to identify and transcribe it.

The text itself (in red in Fig. 1) is written in columns, from top to bottom, then from right to left. Commentaries are embedded within main text, in double columns and smaller font (in purple in Fig. 1).

Sometimes, addition text, either manuscript notes or printed metadata, appears in the margins (in yellow in Fig. 1).

2.2 Quantitative Description of the Dataset and Guidelines

The dataset comprises 327 pages, totaling 13,759 transcribed lines (see Table 1). Annotations have been made using Calfa Vision [28], an online collaborative tool that incorporates active learning strategies. As annotations progress, the tool automatically generates and refines layout and text predictions, specifically flagging the most challenging pages for the model to facilitate targeted corrections.

The goal extended beyond simple transcription; it aimed to generate XML files structured according to TEI (Text Encoding Initiative) standards, incorporating semantic interest areas both at the region and baseline levels within those regions. The key types of elements annotated on the page are detailed in Table 2.

Semantic typing prioritizes MainText and Marginalia_MetaData regions, along with Text, Commentary, and Title lines (see Table 2). Despite the dataset's imbalance in character classes (discussed in part 4.2 below), it is representative of the targeted xylographic production theme.

Table 2. Region Types and corresponding baselines

Type	Zone	Count	Definition
Text Regions			
MainText		296	Main Zone of text
MainText Table Of Content		58	Table of Contents
Marginalia MetaData		492	MetaData informations
Marginalia Caption		49	Caption for an image
Marginalia PageNumber		106	Regions for arabic numbers
Marginalia		23	Additional notes by readers
Image		53	Regions for Illustrations
Image Stamp		27	Regions for Library Stamps
Image Seal		22	Regions for Seals
Text Lines			
Author_name	MainText*	220	
Commentary	MainText*	6633	
MarginaliaLine	Marginalia	89	
Page_Number	Marginalia MetaData	0	
Text	MainText*	4190	
Title	MainText*, Marginalia Metadata and Caption	2093	

In order to ensure the interoperability of the dataset with other common annotation tool or models, we have adopted an annotation by baseline from which polygons are generated. On the other hand, the baseline information is not used in the present experiments. The data is distributed in pageXML format, under Apache 2.0 license.

3 Layout Analysis and Reading Order of Chinese Handwriting

3.1 Simple Columns for Text, Double Columns for Commentaries

The order of the columns is complexified by a layout that mixes main text and exegetical apparatus. In the selected body of texts, the Confucian Classic called *Shijing* 詩經 (Classic of Poems) best exemplifies this complexity. An official version of this anthology of 300 poems was established on imperial order and commented upon by a certain Kong Yingda 孔穎達 (574–648) and his team, who systematically cite three other commentaries: a preface which dates back to the 1st century BC, a system of glosses attributed to a certain Mao 毛, probably also from the 1st century BC, and a sub-commentary by the scholar Zheng Xuan 鄭玄 (127–200). Kong Yingda then adds his own prefaces and comments. Glosses and comments are inserted immediately after the passage or words to which they relate. They are recognizable thanks to their format: instead of occupying the entire width of a column, commentaries are written in double columns, in a font half the size of main text. The same layout applies in most editions, including that of the *Yutai xinyong* (Fig. 1), where we have drawn on the right page two red binding boxes to identify two strings of main text, and purple binding boxes for glosses. A blue binding box corresponds to the title of the next poem. In a large column on the page, each block corresponding to a single category is read before the one below. Also, within each block, baselines read from top to bottom and from right to left.

Several of the texts in our corpus are characterized by this textual mille-feuille inside the columns. However, as the case of the *Classic of Poems* detailed above illustrates, the distinction between the different parts of the text is not trivial: each of the comments points to a certain part of the base text (the one that is just above), each one follows its own discursive modalities, and each dates from a period distinct from the others. The main text also dates to a specific period. In the case of the *Classic of Poems* anthology, each poem was composed between the 11th and 6th centuries BC, although the written version which has reached us was not been established before the 2nd century BC.

Distinction between types and correct sequence of baselines are hence critical.

3.2 Strategies to Establish Reading Order

While layout analysis tasks, including complex ones, have been largely addressed in most common scenarios through targeted fine-tuning (excluding forms and newspapers) using either a basic U-net architecture for segmentation [1,8,19,28] or, more recently, Transformer models [30,32], the challenge of determining the reading order of text regions and lines persists, particularly in the context of historical documents. The reading order represents a semantic challenge that cannot be directly solved using only computer vision techniques. Yet, it remains a critical factor for ensuring the accuracy and relevance of OCR/HTR outputs.

A topological sorting [2] proves adequate for simple document layouts, such as those limited to two columns with marginal annotations, but it encounters limitations with more complex documents. To address this, more advanced approaches have been employed. These include Graph Segmentation without machine learning [29] and the use of a multi-layer perceptron with machine learning [20]. However, these methods primarily capture morphological features, such as the spatial relationships between text regions or lines, rather than semantic attributes. Vision Transformers offer a promising solution, as they integrate both spatial and contextual information [30]. Although they have been tested predominantly on printed documents, training them requires a substantial data volume. This requirement often proves challenging for historical documents, particularly those with limited resources.

The Chinese reading order introduces additional complexity, typically following a hierarchical right-to-left and top-to-bottom pattern. While topological sorting based on centroid positions of text areas is straightforward in most scenarios (as demonstrated by the dataset in this study), applying the same logic to lines within text boxes proves problematic. This is illustrated in Fig. 2, which shows a mix of Text (in red) and Commentary (in purple). Global topological sorting here is significantly hindered by its reliance on the coordinates of the bounding boxes or polygons of the lines. An initial experiment using a simple global sort of line centroids achieved an accuracy of 82.31%. However, this accuracy dropped below 60% with variations in baseline inclination, page orientation, and the model's ability to predict complete baselines in practical scenarios.

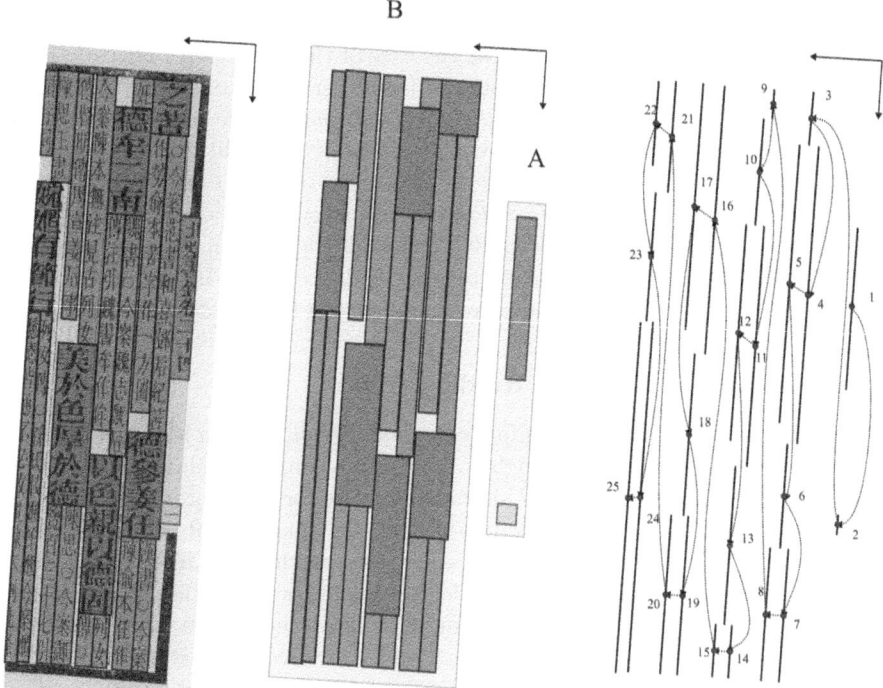

Fig. 2. Reading Order in Chinese Documents: Commentary line in purple, Text line in red, Title in green, and Page number in yellow. Traditional methods using box or baseline centroids often fail to accurately determine reading order in this case. (Color figure online)

The top-bottom and right-left hierarchy often conflicts, as exemplified by the lines labeled 3-4-5, which may be incorrectly ordered as 4-3-5 or 3-5-4, and the lines 11–12, where the shorter line 12 has a higher centroid (see Fig. 2).

To address the challenge of sorting Chinese lines, two main approaches are prevalent in state of the art. The first approach employs a non-global, algorithmic sorting method, which establishes a topological hierarchy starting from pairs of characters and extending to larger groupings, achieving a final accuracy of 97.7%, even on pages that are wavy or misoriented [16]. This method, however, depends on glyph bounding boxes, which are not commonly included in annotations schemes and may be inaccurately defined by recognizers, as perfect alignment with glyphs is not essential for their recognition. The second approach utilizes machine learning, specifically a convolutional neural network (CNN) trained to discern line spacing and character sizes [17], coupled with a multi-layer perceptron (MLP) [19]. This method integrates both global (MLP) and local (CNN) sorting techniques, resulting in a Page Error Rate of 5%, where a page is deemed incorrect if any line is misordered.

3.3 Proposed Method and Results

The overall challenge of determining the reading order is partially mitigated in both scenarios through a local analysis of the relationships between objects. However, the unique characteristics of our dataset – marked by a high variety of document types and a significant mix of semantic line types (see Part 2.2) – complicate the reproducibility of these methods. Additionally, the lack of glyph-level annotations further limits their effectiveness. Consequently, training a multi-layer perceptron (MLP) for global sorting yields an accuracy of only 50%, which is effectively equivalent to random sorting.

In this paper, we propose incorporating the reading order task as an auxiliary component within the layout analysis pipeline (see Fig. 3). Specifically, we apply a localized algorithmic topological sort that efficiently handles dual-column layouts:

- **Horizontal Overlap with Right Priority**: If two lines overlap vertically beyond a certain threshold, the line whose centroid is further to the right is considered to come first. This reflects a reading order from right to left within overlapping lines.
- **Vertical Priority without Overlap**: If two lines do not vertically overlap beyond the threshold, the line whose centroid is higher (lower y-coordinate) is considered to come first. This mimics the top-to-bottom reading order for non-overlapping lines.

We formalize the reading order determination through an overlap_ratio and a sorting function S that takes into account the bounding coordinates of two lines u and v. The function is defined as follows:

$$\text{overlap\_ratio}(u, v) = \frac{|\mathcal{I}_u \cap \mathcal{I}_v|}{|\mathcal{I}_u \cup \mathcal{I}_v|}$$

where:

- $\mathcal{I}_u = [y_{\min,u}, y_{\max,u}]$, and $\mathcal{I}_v = [y_{\min,v}, y_{\max,v}]$,
- $y_{\min,u}$ and $y_{\max,u}$ denote the minimum and maximum y-coordinates of line u's bounding-box (respectively for v).

We have finally, for a given δ threshold :

$$S(u,v) = \begin{cases} 1 & \text{if overlap\_ratio}(u,v) > \delta \text{ and } u_x > v_x \\ 1 & \text{if overlap\_ratio}(u,v) \leq \delta \text{ and } u_y < v_y \\ 0 & \text{otherwise} \end{cases}$$

where u_x, v_x, u_y, and v_y represent the x and y coordinates of the centroids of lines u and v.

The layout analysis process comprises three stages: First, the detection and semantic classification of text regions are performed using a CNN-anchor-based instance segmentation model, more relevant in this under-resourced case than state-of-the-art anchorless approach for Chinese [24,25]. Instance segmentation ensures that closely positioned or similar regions do not merge, a well-know phenomenon with pixel-level semantic segmentation [15], and allows for accurate detection on curved or poorly oriented pages. Second, lines of text are detected and semantically classified, also through instance segmentation. Third, double columns of text are identified using a CNN-based object detection model. We are using a YOLOv8-s model for each step, with default hyperparameters and data augmentation [14].

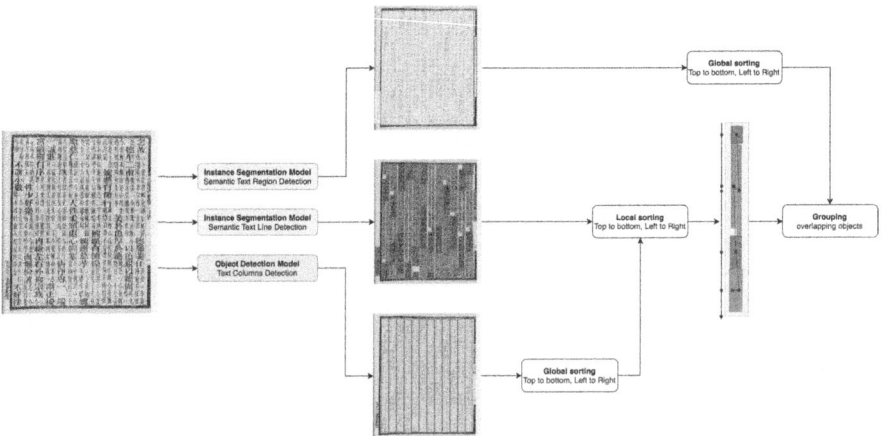

Fig. 3. Pipeline for layout analysis and global/local reading orders

After these three stages, a global sort is applied to both the regions and text columns. The lines within each column are then sorted according to the previously described algorithm. Finally, a global merge of all overlapping objects is executed to generate the page's output (see Fig. 3). While annotation tools commonly utilize baselines for quicker document markup [15,28], our approach opts for direct detection and sorting of lines. This method proves more effective and expedient for managing the complexities of Chinese documents, which often feature dense text layouts where traditional baseline strategies may not be optimal.

Table 3 summarizes the detection results for each task. For text regions, the model achieves an overall precision of 0.95 and an overall recall of 0.93. This paper specifically targets the MainText, MainText TableOfContents, and Marginalia_Metadata regions. Detection and classification are nearly perfect for the first two categories, though there are some failures with MainText TableOfContents. Nevertheless, the distinction between MainText and MainText TableOfContents is ambiguous, primarily due to differences in line indentation within the region. When MainText TableOfContents is not identified, it is classified as MainText, which suffices for the purposes of this study.

The detection of Marginalia_Metadata presents more challenges due to its variable location along the right and left page borders, often complicated by page folds that crop the content. Despite these difficulties, the model still achieves both precision and recall rates around 0.9.

Table 3. Layout analysis results for the three models of detection

	P	R	mAP
TextRegions detection (mask)			
Image	0.952	0.842	0.850
Image_Stamp	0.965	1.000	0.995
MainText	0.993	0.965	0.991
MainText_TableOfContents	0.955	1.000	0.995
Marginalia	0.910	1.000	0.995
Marginalia_Caption	1.000	0.953	0.995
Marginalia_Metadata	0.908	0.884	0.917
Marginalia_PageNumber	0.917	0.793	0.914
Column detection (bbox)			
Column	0.972	0.890	0.937
TextLines detection (mask)			
Author_name	0.419	0.714	0.722
Commentary	0.958	0.944	0.976
MarginaliaLine	0.774	0.977	0.959
Page_Number	0.748	0.676	0.72
Text	0.968	0.979	0.979
Title	0.896	0.855	0.929

Column detection exhibits varying precision and recall, suggesting an overdetection of columns compared to expectations (False Positives). The model was trained exclusively on annotations from MainText and MainText TableOfContents columns (regions critical for maintaining correct reading order). Additionally, many pages at the end of chapters contain extensive blank spaces, which are not fully written. The model also tends to incorrectly tag columns delineated by lines, even though these are not present in the training annotations. While this leads to numerous false positives, they do not adversely affect the overall results, as they typically involve empty text columns or columns outside the target regions.

Line detection yields variable results, with Commentary and Text classes being accurately detected (mean precision of 0.963 and mean recall of 0.961). However, Author_Name and Title classes show greater variability due to their visual similarity to Text in MainText and MainText TableOfContents. These classes are differentiated only by the presence of indentations, which are not included in the annotations (bounding boxes start at the text, not the preceding space). For Title, the ambiguity increases in certain books where indented lines are classified as Text rather than Title, leading to notable misclassifications. Despite these challenges, the lines are in any case detected and sorted correctly, although their classification may be incorrect.

We evaluate reading order accuracy in two ways: Firstly, of the 412 lines assessed, 9 were incorrectly sorted, resulting in an accuracy of 97.81%. Secondly, from an editorial perspective, we consider a page faulty if at least one line is incorrectly sorted, yielding an accuracy of 93%. Current errors appear to be associated with very specific cases in how algorithm handles overlapping lines, indicating a need for further investigation.

4 Text Recognition

Both classical and modern Chinese are written in characters (sinograms) used singly or in combination to form words that are not separated by spaces. The great dictionary of Chinese characters, *Hanyu da zidian*, listed over 60,000 sinograms in 2010 [9]. The challenge this represents for machine recognition is further compounded, as we shall see, by the numerous graphic variants.

4.1 Words, Characters and Glyphs in Ancient Chinese Texts

It is historically attested that certain works were transcribed according to precise stylistic standards, although this constraint does not erase the diversity of hands, as Huang Mingli has shown in a recent article devoted to the court style *guange ti* 館閣體 in the great imperial collection entitled *Siku Quanshu* 四庫全書 (Complete Book of the Four Stores) [13]. In order to better characterize the graphic variety of the annotated corpus, we need to explain what the xylographic reproduction that is most common in imperial China is based on, clarify the notion of graphic variant, and return to the awareness of this variety evidenced in certain ancient writings. The reproduction of text by printing on paper actually began in the 9th century CE, although this method is attested much earlier for the printing of images (notably Buddhist), charms and calendars [5,18,26].

Nevertheless, xylographic plates are derived from handwritten copies, so the results of this printing method do not lead to graphic standardization, as is the case with movable type printing introduced by Johann Gutenberg in 15th-century Europe. The consequences of the technical choice of xylographic printing include irregularity in the transcription of characters, a variety of hands within the same work, and a diversity of writing styles from one edition to the next.

The variety we just described is inherent to manuscript traditions and a priori expected in an HTR project. In the case described, the challenge is all the greater since Chinese language is by definition written using a very large number of characters-the classical Chinese language of the first millennium in particular. The problem becomes that of the annotator (as opposed to the computer scientist) when the question of character variants arises.

A variant (in Chinese *yi ti zi* 異體字) does not correspond to a regular graphic expression of a graphic unit, like an 'a' whose transcription varies according to the font or case of the character. The equivalent of such variety is writing styles (sometimes labelled calligraphic styles), and called *ti* 體 in Chinese. By contrast, a variant corresponds to a graphic expression of a script character that has ceased to be used, that is considered exceptional or even faulty, or that has been defined to enable the identification (by equivalence) of an otherwise taboo character[9] in an edition.

These variants stand as challenges for the transcriber, as they are not normally part of the unicode character set. The subject is made all the more thorny by the fact that a number of these variants exist in unicode. In other words, given that many rare characters (and therefore poorly known to annotators) are affected by these variant problems, it is difficult:

1. to identify the variant as such,
2. to transcribe it using the standard version of the character to which it corresponds.

The choice was made to standardize the transcription of the corpus, so as to avoid variant confusion as much as possible, and to limit, so to speak, the overall number of different characters.

4.2 Recognition Results

Chinese HTR has predominantly advanced in online recognition and contemporary datasets, as seen in recent ICDAR publications [23]. CRNN+CTC models typically deliver good results on Chinese texts [12], although they demand large data volumes and struggle with generalization [3]. The primary challenge is

---

[9] The personal name of the emperor was taboo during his lifetime, meaning that the characters that make up his personal cannot be written in any edition conceived during his lifetime and need to be replaced by substitute characters.

the extensive number of classes (several thousands) that need to be recognized. Transformers present a viable alternative, offering excellent adaptability and handling of zero-shot learning scenarios [31], yet their effectiveness is limited on historical documents due to data scarcity. The SVTR-net (Single Visual Model) shows the best outcomes even for historical texts but requires training on several million characters [6,17]. While transfer learning could be a relevant approach for our dataset, which features 5,581 unique characters – meaning 30.46% of which are represented only once, it is not currently feasible. Conversely, some characters are highly represented in our dataset, highlighting the imbalance. For instance, characters such as the following ones are particularly prevalent: 之 (2,239 samples), 也 (1,552), 日 (1,549), 不 (942), 以 (895), 一 (889), 而 (769), 有 (765), 十 (742) and 為 (737)-as opposed to 佯, 棰, 帕, 伉, and 儡 that have only one sample.

Existing models fail to perform adequately on our out-of-domain corpus, typically achieving no more than 40% weighted-accuracy [3]. This limitation is particularly critical in few-shot, single-shot, or zero-shot learning contexts where many characters may never be encountered during training, making existing models unsuitable for our needs.

To benchmark this dataset, we employ a word-based CRNN architecture, which provides greater contextual understanding for recognition [28]. Addressing the data scarcity issue, we implement qualitative data augmentation using CycleGAN [33], previously proven effective for Chinese text [4]. CycleGAN has been experimented to transfer handwriting styles to printed text for HTR data creation in Latin scripts [27], and authors have shown its limitations such as scale discrepancies and the absence of a recognizer to ensure real readability of the generated image, like the one used in ScrabbleGAN [7,27]. The GAN approach is applied in a highly constrained setting, using cropped images of characters with a simple handwritten effect mapped onto printed characters. This reduces GAN instability, although no control experiments were conducted.

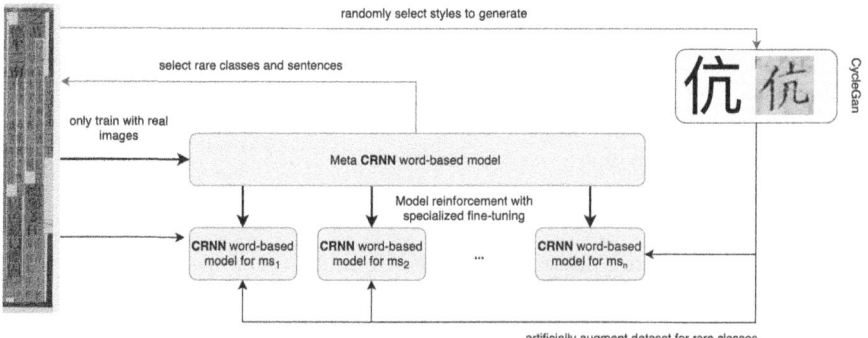

Fig. 4. Pipeline for training recognition model

For our purposes, we selectively generate single characters during training to enhance the representation of underrepresented classes. To maximize context in vocabulary, word sequences, and graphic variations, we implement a dual-learning approach. Initially, a model is trained using the entire dataset comprised solely of real data. This model then constitutes a base model for fine-tuning a specialized one tailored to the target manuscript. If needed, CycleGAN is utilized to augment the dataset with rare characters identified in the first training phase (see Fig. 4). We ensure consistent data distribution across training, validation, and testing phases. Although this strategy results in some over-fitting, it is deemed acceptable for processing a specific target manuscript. Table 4 summarizes recognition results for each manuscript.

Table 4. Recognition results

Manuscript	N°	Accuracy
Li Wenxuan	A-1	99.38 (±1.2)
Liuchen Wenxuan	A-2	98.84 (±1.8)
Yutai	A-3	98.52 (±1.2)
Tangshi	A-4	99.25 (±1.8)
Beitang	S-1	98.76 (±1.8)
Bowu zhi	S-2	99.18 (±1.8)
Chuxue	S-3	97.57 (±1.7)
Erya	S-4	96.57 (±0.4)
Maoshi shu	S-5	98.42 (±1.8)
Yiwen	S-6	98.72 (±1.7)
Zhibuzu	S7	98.70 (±1.8)
Shiwen leiju	T-1	97.47 (±4.5)
Qimin yaoshu	T-2	99.35 (±2.8)
Xinzhai	T-3	97.61 (±3.2)

The results demonstrate that this strategy can yield high accuracy, even with small datasets; three manuscripts achieved around 97% accuracy.

If we look in more detail, we find the average confusion in character recognition reaches a peak for characters with 10–20 samples. However, for this specific case, the situation varies a lot depending on the manuscript taken into consideration. A further inquiry on these cases would be needed to explain and hopefully improve the results. On the other hand, characters with fewer than 10 samples show similar recognition rates to those with between 100 and 400 samples. Not that both situations are exactly similar: confusion rates appear more stable for characters with large samples in most manuscripts, when they tend to vary more in one-shot learning situations. The recognition accuracy for unknown characters, including those that were artificially generated, stands at 86.21%. Even though part of the explanation may lie in the fact that these characters are infrequent in both the training phase and during inference, these results show that the strategy developed in this project is efficient when it comes to scarce characters.

To illustrate the real-world applicability of the models, we also provide adjusted accuracy figures ($\pm 1.8\%$) that account for predictions following uncontrolled layout detection. However, in real-world scenarios, the accuracy is notably reduced by issues such as Marginalia_MetaData texts being split in half, resulting in less than 60% accuracy. In the MainText category, the primary errors would stem from incomplete or noisy segmentation of lines, often caused by line polygons that cut through characters.

5 Conclusion

The dataset used in these experiments presents substantial challenges, aligning with the project's ambitious objectives. This dataset has been made publicly available, providing a base and benchmark for developing versatile HTR models for Chinese historical documents. While results for reading order need further enhancement, a local algorithmic approach proves adequate for the task when supplemented by additional semantic zone detection, reaching equivalent or better results than ML-based approaches but without training. Analysis of Chinese layouts is generally effective but lacks robustness in high-density text situations. Future plans include employing Transformer-based layout analysis on this dataset. Text recognition leveraged a GAN-supported CRNN, achieving an average accuracy of 98.45% ($\pm 1.9\%$). A refined analysis of the impact of character frequency in the dataset on character recognition rates shall contribute to a better understanding of the choices to be made at the stage of dataset selection. Finally, while the models are not adaptable to out-of-domain applications, they meet the specific editorial requirements of the project effectively. Given the great variety of the dataset, these models stand as great candidates to be tested and fine-tuned for ancient Chinese manuscript recognition.

Acknowledgments. The CHI-KNOW-PO Project was funded by the University of Strasbourg Institute for Advanced Studies (USIAS) and the COLLEX-Persée. It has been conducted in collaboration with three libraries in France, namely, the Bibliothèque universitaire des langues et civilisations (BULAC, Paris), the Bibliothèque nationale et universitaire de Strasbourg (BNU, Strasbourg), and the Bibliothèque de l'institut des hautes études chinoises at the Collège de France (BIHEC, Paris). The Calfa start-up was in charge of developing HTR models.

Data Availability Statement. Webpage of the project: https://www.collexpersee.eu/projet/chi-know-po-corpus/ ; Gitlab of the project: https://chi-know-po.gitpages.huma-num.fr/ ; Dataset: https://github.com/calfa-co/chi-know-po

Disclosure of Interests. The authors have no competing interests to declare that are relevant to the content of this article.

References

1. Boillet, M., Kermorvant, C., Paquet, T.: Multiple document datasets pre-training improves text line detection with deep neural networks. In: 2020 25th International Conference on Pattern Recognition (ICPR), pp. 2134–2141. IEEE (2021)
2. Breuel, T.M.: High performance document layout analysis. In: Proceedings of the Symposium on Document Image Understanding Technology, vol. 5 (2003)
3. Brisson, C., Constant, F., Bui, M.: Chinese historical documents automatic transcription (CHAT) models (2023). https://doi.org/10.5281/zenodo.8383732
4. Chang, B., Zhang, Q., Pan, S., Meng, L.: Generating handwritten Chinese characters using cyclegan. In: 2018 IEEE Winter Conference on Applications of Computer Vision (WACV), pp. 199–207. IEEE (2018)
5. Drège, J.P.: Le livre manuscrit et les débuts de la xylographie. In: Le livre et l'imprimerie en Extrême-Orient et en Asie du Sud. Société des bibliophiles de Guyenne (1986)
6. Du, Y., et al.: Svtr: scene text recognition with a single visual model. In: Proceedings of the Thirty-first International Joint Conference on Artificial Intelligence (IJCAI-31) (2022). arXiv preprint arXiv:2205.00159
7. Fogel, S., Averbuch-Elor, H., Cohen, S., Mazor, S., Litman, R.: Scrabblegan: semi-supervised varying length handwritten text generation. In: The IEEE Conference on Computer Vision and Pattern Recognition (CVPR) (2020)
8. Grüning, T., Leifert, G., Strauß, T., Michael, J., Labahn, R.: A two-stage method for text line detection in historical documents. Int. J. Doc. Anal. Recogn. (IJDAR) **22**(3), 285–302 (2019)
9. Han yu da zi dian bian ji wei yuan hui (The Editorial Committee of the Large dictionary of Chinese characters): 'Hanyu da zidian' (Large dictionary of Chinese characters). Sichuan cishu chubanshe (2010)
10. Henke, K., Arnold, M.: Jing bao ground truth - text block crops and annotations (2023). https://doi.org/10.11588/data/PVYWKB
11. Henke, K., Arnold, M.: Language model assisted OCR classification for republican Chinese newspaper text. J. Digit. Arch. Digit. Hum. **11**, 1–19 (2023)
12. Hu, S., Wang, Q., Huang, K., Wen, M., Coenen, F.: Retrieval-based language model adaptation for handwritten Chinese text recognition. IJDAR **26**(2), 109–119 (2023)
13. Huang, M.L.:'siku quanshu' tenglu shufa fengmao fenlei chutan - yi wenyuange ben wei zhu (manuscript calligraphy styles of siku quanshu based on wenyuan pavilion version). Zhongguo xueshu niankan **40**(2), 27–57 (2018)
14. Jocher, G., Chaurasia, A., Qiu, J.: Ultralytics YOLO, January 2023. https://github.com/ultralytics/ultralytics
15. Kiessling, B., Tissot, R., Stokes, P., Ezra, D.S.B.: Escriptorium: an open source platform for historical document analysis. In: 2019 International Conference on Document Analysis and Recognition Workshops (ICDARW), vol. 2, pp. 19–19. IEEE (2019)
16. Lee, A., Yu, H., Min, G.: An algorithm of line segmentation and reading order sorting based on adjacent character detection: a post-processing of ocr for digitization of chinese historical texts. J. Cult. Herit. **67**, 80–91 (2024)
17. Ma, H.Y., Huang, H.H., Liu, C.L.: Reading between the lines: image-based order detection in ocr for chinese historical documents. In: Proceedings of the AAAI Conference on Artificial Intelligence, vol. 38, pp. 23808–23810 (2024)
18. Pelliot, P.: Les débuts de l'imprimerie en Chine. Imprimerie Nationale - Adrien Maisonneuve (1953)

19. Quirós, L.: Multi-task handwritten document layout analysis. In: Proceedings of the Twenty-Seventh International Joint Conference on Artificial Intelligence (IJCAI-18), pp. 1057-1063 (2018). arXiv preprint arXiv:1806.08852
20. Quirós, L., Vidal, E.: Reading order detection on handwritten documents. Neural Comput. Appl. **34**(12), 9593–9611 (2022)
21. Shi, B., et al.: ICDAR2017 competition on reading chinese text in the wild (RCTW-17). In: 2017 14th IAPR International Conference on Document Analysis and Recognition (ICDAR), vol. 1, pp. 1429–1434. IEEE (2017)
22. Sturgeon, D.: Large-scale optical character recognition of pre-modern chinese texts. Int. J. Buddhist Thought Culture **28**, 11–44 (2018)
23. Su, T., Zhang, T., Guan, D.: Corpus-based hit-mw database for offline recognition of general-purpose chinese handwritten text. IJDAR **10**, 27–38 (2007)
24. Tang, C.W., Liu, C.L., Chiu, P.S.: Hrcenternet: an anchorless approach to chinese character segmentation in historical documents. In: 2020 IEEE International Conference on Big Data (Big Data), pp. 1924–1930. IEEE (2020)
25. Tang, C.-W., Liu, C.-L., Chiu, P.-S.: HRRegionNet: Chinese Character Segmentation in Historical Documents with Regional Awareness. In: Lladós, J., Lopresti, D., Uchida, S. (eds.) ICDAR 2021, Part IV. LNCS, vol. 12824, pp. 3–17. Springer, Cham (2021). https://doi.org/10.1007/978-3-030-86337-1_1
26. Twitchett, D.C.: Printing and publishing in medieval China. Frederic C, Beil (1983)
27. Vidal-Gorène, C., Camps, J.B., Clérice, T.: Synthetic lines from historical manuscripts: an experiment using GAN and style transfer. In: Foresti, G.L., Fusiello, A., Hancock, E. (eds.) Image Analysis and Processing - ICIAP 2023 Workshops, ICIAP 2023, LNCS, vol. 14366, pp. 477–488. Springer, Cham (2024). https://doi.org/10.1007/978-3-031-51026-7_40
28. Vidal-Gorène, C., Dupin, B., Decours-Perez, A., Riccioli, T.: A modular and automated annotation platform for handwritings: evaluation on under-resourced languages. In: Lladós, J., Lopresti, D., Uchida, S. (eds.) ICDAR 2021, Part III. LNCS, vol. 12823, pp. 507–522. Springer, Cham (2021). https://doi.org/10.1007/978-3-030-86334-0_33
29. Wang, R., Fujii, Y., Bissacco, A.: Text reading order in uncontrolled conditions by sparse graph segmentation. In: Fink, G.A., Jain, R., Kise, K., Zanibbi, R. (eds.) Document Analysis and Recognition - ICDAR 2023, ICDAR 2023, LNCS, vol. 14192, pp. 3–21. Springer, Cham (2023). https://doi.org/10.1007/978-3-031-41731-3_1
30. Wang, Z., Xu, Y., Cui, L., Shang, J., Wei, F.: Layoutreader: pre-training of text and layout for reading order detection. In: Proceedings of the 2021 Conference on Empirical Methods in Natural Language Processing, pp. 4735–4744 (2021). https://doi.org/10.18653/v1/2021.emnlp-main.389, https://aclanthology.org/2021.emnlp-main.389, arXiv preprint arXiv:2108.11591
31. Yu, H., Wang, X., Li, B., Xue, X.: Chinese text recognition with a pre-trained clip-like model through image-ids aligning. In: Proceedings of the IEEE/CVF International Conference on Computer Vision, pp. 11943–11952 (2023)
32. Zhang, N., et al.: M2doc: a multi-modal fusion approach for document layout analysis. In: Proceedings of the AAAI Conference on Artificial Intelligence, vol. 38, pp. 7233–7241 (2024)
33. Zhu, J.Y., Park, T., Isola, P., Efros, A.A.: Unpaired image-to-image translation using cycle-consistent adversarial networks. In: 2017 IEEE International Conference on Computer Vision (ICCV) (2017)

Mind the Gap: Analyzing Lacunae with Transformer-Based Transcription

Jaydeep Borkar(✉) and David A. Smith

Khoury College of Computer Sciences, Northeastern University,
Boston,
MA 02115, USA
borkar.j@northeastern.edu, dasmith@ccs.neu.edu

Abstract. Historical documents frequently suffer from damage and inconsistencies, including missing or illegible text resulting from issues such as holes, ink problems, and storage damage. These missing portions or gaps are referred to as lacunae. In this study, we employ transformer-based optical character recognition (OCR) models trained on synthetic data containing lacunae in a supervised manner. We demonstrate their effectiveness in detecting and restoring lacunae, achieving a success rate of 65%, compared to a base model lacking knowledge of lacunae, which achieves only 5% restoration. Additionally, we investigate the mechanistic properties of the model, such as the log probability of transcription, which can identify lacunae and other errors (e.g., mistranscriptions due to complex writing or ink issues) in line images without directly inspecting the image. This capability could be valuable for scholars seeking to distinguish images containing lacunae or errors from clean ones. Although we explore the potential of attention mechanisms in flagging lacunae and transcription errors, our findings suggest it is not a significant factor. Our work highlights a promising direction in utilizing transformer-based OCR models for restoring or analyzing damaged historical documents.

Keywords: Optical Character Recognition · Transformers · Lacuna

1 Introduction

Many historical documents have reached us in incomplete form. Besides the larger losses of missing leaves or inscribed surfaces, existing pages and other carriers of writing often exhibit lacunae, gaps at the edge or in the middle of lines, or missing lines in the middle of a text. Moreover, when working with digital surrogates for historical documents, we may encounter gaps introduced by the imaging process, such as overly dark microfilm or poorly cropped photographs. Paleography, papyrology, and epigraphy have developed standards, such as the Leiden conventions, for distinguishing readable text, partial letters, abbreviated forms, and restored lacunae [7].

Building on progress in handwritten text recognition (HTR) [11] and language modeling [5], researchers have employed language technologies to good effect in the traditional philological task of hypothesizing readings for these lacunae *when their locations are known* [2,4]. If asked to transcribe a new document, however, these restoration models do not know where any gaps in them might be. Furthermore, as language technologies become more powerful, transcription models may suggest readings that are not supported by input document images (or, similarly, sound recordings [9]).

This paper considers two related inference problems that arise when HTR models are confronted with images of a document that may contain lacunae. First, is the HTR model capable of accurately filling the gap left by the absence of visual evidence of writing? Second, does the model provide information about the presence of lacunae, so that users are aware of which parts of a transcript are inferred from context alone? We address the first question by measuring HTR transcription accuracy on lines with and without lacunae, evaluating models trained with various proportions of lacunae. We address the second question by training models to predict the presence of lacunae and other errors in text lines at inference time.

As in many other areas of natural language processing, transformer architectures are being utilized in OCR. We perform experimental evaluations with the TrOCR model, which combines a pre-trained vision transformer with a pre-trained language model decoder [10]. Besides state-of-the-art performance, transformer architectures employ an attention mechanism, which provides us with a way to probe the relationship between visual evidence and transcription output. To control the prevalence of lacunae experimentally, we identify the presence of individual characters in the training and test data using OCR and randomly assign some proportion of them to be missing.

This study reaches the following main findings:

- Transformer models like TrOCR that are pre-trained on clean text struggle to transfer their knowledge effectively for inferring (the likely content of) lacunae in line images.
- Adding examples with lacunae to TrOCR's supervised training significantly enhances its rate of lacuna restoration from 5.6% to 65.85%.
- Using TrOCR's log probability of an HTR transcript as a predictor in a logistic regression model correctly flags lines containing lacunae 53% of the time and images with other errors 84% of the time.
- Using a sample of the model's attention weights to predict lacunae in input images and mistranscriptions in the output achieves accuracy not significantly higher than what was already reached using the log probability of the transcription. Nonetheless, we believe further experiments probing HTR models' representations of their input would be useful.

After reviewing related work (Sect. 2), we describe the construction of the training and test data for our study and the inference procedures used in our experiments (Sect. 3). We then describe the experimental setup (Sect. 4) and discuss the results (Sect. 5). We conclude with a discussion of future directions for research on lacunose manuscripts (Sect. 6).

2 Related Work

Efforts have been made to restore lacunae through a pretrain-finetune framework [16]. In this approach, authors utilized self-supervised contrastive loss training, followed by supervised fine-tuning. This method showed significant improvement in the restoration of lacunae within printed and handwritten documents, both in English and Arabic languages. There have been numerous works in NLP in the domain of masked pretraining where the objective is to learn the masked representations from unlabeled data [5]. In the speech domain, Baevski et al. proposed wav2vec 2.0 framework that masks the speech input and solves a contrastive task over quantized speech representations by using self-supervised contrastive loss during pre-training and Connectionist Temporal Classification (CTC) loss during fine-tuning [3]. Damage and inconsistencies in historical documents, such as inking issues, often pose challenges when using modern OCR models for transcription [1]. Several OCR post-correction models have been proposed to correct the transcription of OCR models when dealing with historical documents [6,8], including some for endangered languages [14,15].

Transformer-based models have been widely used in language modeling tasks [5,12,13] and in optical character recognition to transcribe the text present in the image [10]. Transformer-based OCR models, such as TrOCR, use an encoder-decoder framework. In this framework, a pre-trained image transformer serves as the encoder, while a pre-trained text transformer acts as the decoder. TrOCR has demonstrated superior performance over current state-of-the-art models across various domains, including printed, handwritten, and scene text recognition tasks.

3 Approach

3.1 Dataset Creation

In this study, we create synthetic lacunae by utilizing line images from the IAM handwriting database[1], replicating real-world scenarios. We begin by extracting all bounding boxes along with their coordinates from the line image using Pytesseract[2]. Next, we randomly select a bounding box from the list and apply binary thresholding to that region with a threshold value of 10, rendering it completely white to create the lacuna. Gaps generated in this way resemble the lacunae introduced into binarized microfilm images.

Subsequently, we binarize the remaining portions of the line image to improve quality for optical character recognition. We explore two approaches for binarizing line images: static thresholding, which applies the same threshold to all

---

[1] https://fki.tic.heia-fr.ch/databases/iam-handwriting-database.
[2] https://github.com/tesseract-ocr/tesseract.

images, and adaptive thresholding, which dynamically adjusts the threshold value for each image based on its pixel distribution and layout. Our findings reveal that static thresholding adversely affects the visual quality of non-lacuna characters. Therefore, we opt for adaptive thresholding, which preserves the visual integrity of non-lacuna characters. Consequently, we utilize adaptively thresholded binarized images for lacuna creation in our experiments.

We also process the ground-truth labels to address certain irregularities. Initially, characters of words that were originally separated by white spaces are recombined (e.g., "B B C" becomes "BBC"). Additionally, contractions that were inconsistently tokenized from the original data, separating them from the words, are reattached to the respective words (e.g., "He'll" becomes "He'll", "They've" becomes "They've"). Figure 1 shows some samples of our line images containing lacunae.

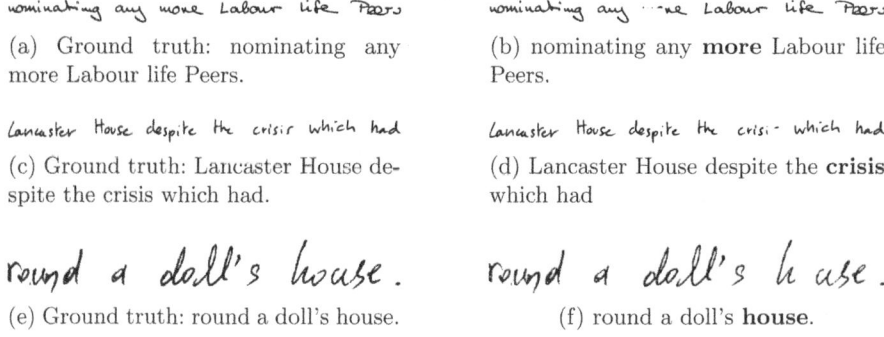

(a) Ground truth: nominating any more Labour life Peers.

(b) nominating any **more** Labour life Peers.

(c) Ground truth: Lancaster House despite the crisis which had.

(d) Lancaster House despite the **crisis** which had

(e) Ground truth: round a doll's house.

(f) round a doll's **house**.

Fig. 1. Line images (b, d, f) containing lacuna highlighted in bold.

3.2 Inference and Evaluation

To evaluate the performance of models on our datasets, we use the Character Error Rate (CER), a metric that is widely used to evaluate the effectiveness of optical character recognition. It measures the rate of character-level errors in the OCR output compared to the ground truth text. CER is given by

$$\text{CER} = \frac{N_{\text{insertions}} + N_{\text{deletions}} + N_{\text{substitutions}}}{N}$$

where N is the total number of characters in the ground truth, $N_{\text{insertions}}$ is the number of characters incorrectly inserted by the model, $N_{\text{deletions}}$ is the number of characters incorrectly deleted by the model, and $N_{\text{substitutions}}$ is the number of characters incorrectly substituted by the model. Let X represent the line image, S denote its text sequence, n signify the total number of words in S, L denote the lacuna token, and $\hat{L}$ symbolize the model's predicted lacuna token. With $S = (w_1, w_2, ..., w_n)$ as the text sequence comprising n words, we assess whether $\hat{L} = L$ when L appears at position n_i in S.

3.3 Baseline: Effects of Lacunae and Complex Writing on Log Probabilities

We aim to investigate the impact of lacunae and other factors, such as characters or words written in an unusual manner, on the log probability of transcription tokens. Log probability is a measure used to quantify the likelihood of a word or sequence of words occurring given some context. In the context of OCR, it represents the probability of predicting a new token given previously transcribed tokens. Let $w_1, w_2, ..., w_n$ be a sequence of n tokens in the transcription, $w_{1:(i-1)}$ be the sequence of tokens from the first token up to w_i, $P(w_i|w_{1:(i-1)})$ be the conditional probability of observing token w_i given the context $w_{1:(i-1)}$. The log probability of a token w_i given its context $w_{1:(i-1)}$ is:

$$\text{Log Probability} = \log P(w_i|w_{1:(i-1)})$$

Our hypothesis suggests that the log probability would decrease for tokens affected by lacunae, as the model would exhibit uncertainty in its interpretation. Similarly, this hypothesis extends to tokens written in a complex manner, where the model may also struggle to accurately predict them thus resulting in lower log probabilities. To test our hypothesis, we train a logistic regression model using the minimum of the log probability ($min(logprobability)$) as a feature. The $min(logprobability)$ represents the lowest log probability observed among all transcription tokens. Let $P_{\min}$ be the minimum of all log probabilities in the transcription tokens, then the logistic regression model can be represented as

$$\text{Logistic Regression: } y = \sigma(\beta_0 + \beta_1 \cdot P_{\min})$$

where y represents the predicted outcome (e.g., lacuna or error detection), β_0 is the intercept, β_1 is the coefficient associated with the feature $P_{\min}$, and σ is the logistic function. We report our findings in Sect. 5.3.

3.4 Hypothesis: Can Attention Detect Lacunae and Complex Writing?

In models such as TrOCR, cross-attention refers to the mechanism by which the models attend to different parts of the input line image while generating the output transcription. Cross-attention helps the model focus on the relevant regions of the input image at each step of the transcription process, enabling it to capture any contextual dependencies for accurate optical character recognition. Given an input image I consisting of a sequence of visual features $\{v_1, v_2, ..., v_n\}$ representing different regions in the image, and a target transcription T consisting of a sequence of tokens $\{t_1, t_2, ..., t_m\}$ representing the characters in the text, the cross-attention mechanism computes a set of attention scores $\{\alpha_1, \alpha_2, ..., \alpha_n\}$, where each attention score α_i indicates the relevance of the ith visual feature to the generation of the current token t_j in the transcription.

We study a possible hypothesis of identifying any attention patterns in the cross-attention matrix to visually identify lacunae and complex writing. Our hypothesis suggests that when the model processes lacuna characters or complex handwriting lacking sufficient visual evidence, the model may allocate greater attention scores to other regions within the line image that provide clearer visual cues. Further, it can utilize this contextual information to infer the identity of the lacuna or other characters written in a complex way. To test this hypothesis, we train a logistic regression model using two features: $min(logprobability)$ (as in Sect. 3.3) and $attentionentropy$. $attentionentropy$ is a measure used to quantify the uncertainty or randomness in the distribution of attention scores. Entropy H for an attention distribution p can be given as

$$\text{Entropy } H(p) = -\sum_{i=1}^{n} p_i \log(p_i)$$

We investigate whether the attention entropy feature has a greater or lesser influence compared to our baseline log probability feature in identifying lacunas and characters with complex handwriting. We report our results related to attention in Sect. 5.3. The reason we train separate logistic regression models for detecting lacuna and other errors is that sometimes the lacuna characters get higher log probability and lower uniform attention when transcribed correctly. Conversely, we see higher attention and lower log probability in the scenario where non-lacuna tokens get mistranscribed due to other errors. This phenomenon makes it a hard problem to solve and thus we decide to train separate models to disentangle these cases. For instance, while transcribing the line image from Fig. 1(f), we find that the model correctly transcribes the lacuna token *house* and gives higher log probability (Fig. 2a) and lower uniform attention (Fig. 2b). Whereas for the non-lacuna tokens that are mistranscribed (such as *record* and *ball*), we see lower log probability and higher non-uniform attention for them.

3.5 Training TrOCR in Leiden Conventions

The Leiden Conventions are adopted in many subfields of paleography to indicate the uncertainty of, and deviation from, the visual evidence for a transcription [7]. Under these conventions, missing or uncertain information, such as illegible characters or words, should be denoted by enclosing them within square brackets. Characters transcribed with partial evidence should have dots placed under them. For instance, using the Leiden conventions, the transcription for the line image in Fig. 1(f) can be written as *round a doll's h[o]use*.

We experimented with training TrOCR using the Leiden conventions in the ground truth transcriptions of line images. We annotated the transcriptions of 15% of the line images containing a lacuna in our training data by enclosing

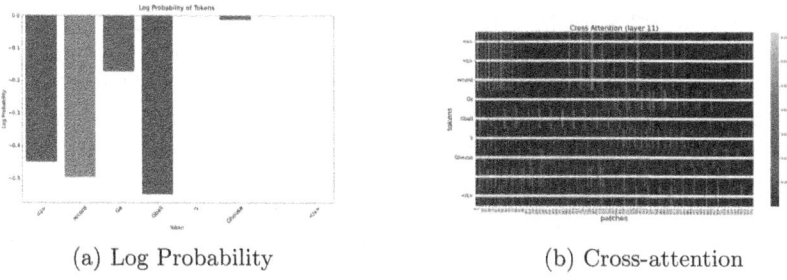

(a) Log Probability (b) Cross-attention

Fig. 2. Log probabilities and cross-attention for the transcripted tokens from Fig. 1(f) with the following ground truth transcription: round a doll's house.

the lacuna token within square brackets. We used a customized loss function to penalize the model for failing to generate square brackets in positions where it should enclose the lacuna within square brackets. This approach aimed to train the model to recognize and appropriately handle lacunae within the transcriptions using the Leiden conventions. Let L be the original (cross-entropy) loss function. The custom weighted loss function L_{weighted} is defined as:

$$L_{\text{weighted}} = L \times (n_{\text{brackets}} + 1)$$

where $(n_{\text{brackets}}+1)$ is the weight and n_{brackets} is the number of occurrences of square brackets [] in the ground truth transcriptions. This equation ensures that the loss for each sample is multiplied by a weight that is proportional to the number of square brackets in its corresponding ground truth transcription. Samples with more square brackets will have higher weights, encouraging the model to learn the convention of generating square brackets around lacuna tokens.

4 Experiments

We use the Aachen's partition of the IAM Handwriting dataset[3] where the training data has 6161 lines from 747 documents, validation data has 966 lines from 115 forms, and test data has 2915 lines from 336 forms. We generate lacuna versions corresponding to all line images in our dataset. Next, we train the following variations of models using the TrOCR base model (334M parameters).

1. Model trained solely on all binarized non-lacuna images (total 6161 images).
2. Model trained on all binarized images and their corresponding lacuna variations (total 6161*2 images).
3. Model trained on all binarized images in addition to 30% randomly selected lacuna images (total 8009 images).
4. Model trained on all binarized images supplemented with 15% randomly selected lacuna images (total 7085 images).

---
[3] https://github.com/jpuigcerver/Laia/tree/master/egs/iam.

The first model is trained to assess whether the Transformer-based model handles lacunae without exposure to any lacuna images. TrOCR is pre-trained without any lacuna variations, so we were interested in evaluating its ability to leverage pre-training knowledge for inferring lacunae. Furthermore, we experiment with different proportions of lacuna images in the training data to observe their impact on model performance. We use the following hyperparameters for training: *learning rate* of 2e-5, *weight decay* of 0.0001, *batch size* of 4, and *AdamW* optimizer. We train our models for 20 epochs on four A100 40GB GPUs and pick the best model (with the lowest validation character error rate).

We examine the attention patterns and log probabilities specifically from the last layer (layer 12) and the first head. The TrOCR base model comprises 12 layers, with each layer containing 16 attention heads. However, in this study, we concentrate solely on the final layer and a single head. While we acknowledge the potential benefits of conducting a comprehensive analysis across all layers and heads, we opt for this focused approach for the current investigation.

5 Results

5.1 Visual Examples: Lacuna Transcription Improvement

In this section, we present images containing lacunae that were initially mistranscribed by the model trained only on non-lacuna images. However, our trained model, which is specifically trained on images with lacunae, successfully transcribes them accurately. We observe in Figs. 3 and 5 that multiple characters from a word form a lacuna, while in Fig. 4, a single-character lacuna is evident. Remarkably, the model trained on lacuna images accurately transcribes all these variations.

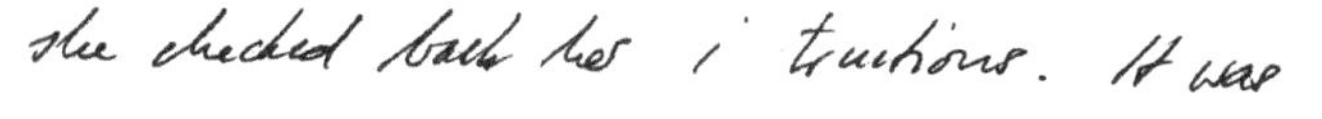

Ground truth: she checked back her instructions. It was
Non-lacuna model: the checked back her 1 truth. It was
lacuna model: she checked back her instructions. It was

Fig. 3. Line image with a lacuna in *instructions*. Errors are highlighted in red. (Color figure online)

5.2 Performance Comparison of Our Various Models

Model Trained only on Binarized Non-lacuna Images : The model trained only on clean binarized line images gives a validation CER of 7.46%, CER of 11.49% on the test set of non-lacuna images, and CER of 13.26% on a mixture of binarized non-lacuna and lacuna line images. Table 1a shows the model's performance in predicting both the lacuna and non-lacuna characters

one of its most cla sic features to the Turkish rule under

Ground truth: one of its most classic features to the Turkish rule under
Non-lacuna model: one of its most dc pictures to the Turkish rule under
lacuna model: one of its most classic features to the Turkish rule under

Fig. 4. Line image with a lacuna in *classic*. Errors are highlighted in red. (Color figure online)

And the sequence in the text is 'fferent than

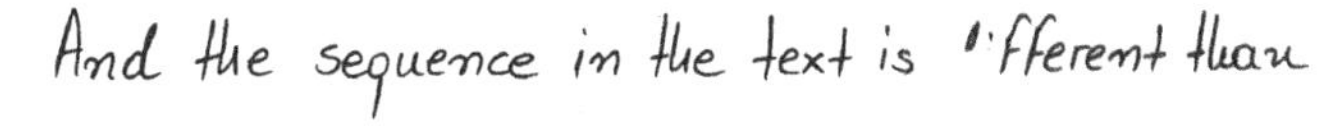

Ground truth: And the sequence in the text is different than
Non-lacuna model: And the sequence in the text is " f frequent than
lacuna model: And the sequence in the text is different than

Fig. 5. Line image with a lacuna in *different*. Errors are highlighted in red. (Color figure online)

in the dataset. We observe that the model accurately transcribes non-lacuna characters 77.82% of the time, while it achieves correct transcriptions for lacuna characters only 5.6% of the time. One potential explanation could be that during pre-training, the TrOCR model may not have encountered any text instances containing missing characters similar to lacunae.

Model Trained on All Binarized Images and Their Corresponding Lacuna Variations : The model trained on clean images and their lacuna variations yields a validation CER of 10% and test CER of 13.75% on a mixture of non-lacuna and lacuna images. Table 1b shows the model's performance in predicting both the lacuna and non-lacuna characters in the dataset. We observe a notable enhancement in model performance when training on lacuna images, increasing accuracy from 5.6% to 65.85%. Specifically, the trained model accurately identifies 1844 out of 2800 lacuna characters, a substantial improvement compared to the baseline model's mere 157 correct predictions. Here, we can assert that training the model on labeled lacuna data teaches it to correctly recognize and restore any lacunae in the text. Another noteworthy observation is that training on lacuna images has a slight impact on the model's accuracy with clean, non-lacuna characters, evidenced by a decrease from 77.82% to 75.38%. This tradeoff implies that while there's an improvement in recognizing lacuna characters, there's a minor compromise in accuracy for non-lacuna characters.

Model Trained on All Binarized Images in Addition to 30% Randomly Selected Lacuna Images : The model trained on 30% lacuna images gives a validation CER of 8.71% and a test CER of 13.04% on a mixture of lacuna and non-lacuna images. Table 1c shows the model's performance in predicting both

Table 1. Confusion matrix for predicting (non-)lacuna characters

(a) Only non-lacuna training

	Lacuna	Non-Lacuna
Correct	5.6% (157/2800)	77.82% (17969/23089)
Incorrect	94.4% (2643/2800)	22.18% (5120/23089)

(b) Non-lacuna + 100% lacuna training

	Lacuna	Non-Lacuna
Correct	65.85% (1844/2800)	75.38% (17406/23089)
Incorrect	34.15% (956/2800)	24.62% (5683/23089)

(c) Non-lacuna + 30% lacuna training

	Lacuna	Non-Lacuna
Correct	62.25% (1743/2800)	76.95% (17767/23089)
Incorrect	37.75% (1057/2800)	23.05% (5322/23089)

(d) Non-lacuna + 15% lacuna training

	Lacuna	Non-Lacuna
Correct	60.14% (1684/2800)	80.35% (18554/23089)
Incorrect	39.86% (1116/2800)	19.65% (4535/23089)

the lacuna and non-lacuna characters in the dataset. We observe a decline in the model's ability to predict lacuna characters compared to when it's trained on 100% of lacuna images. However, its accuracy in predicting non-lacuna characters increases, leading to a lower overall test CER.

Model Trained on All Binarized Images in Addition to 15% Randomly Selected Lacuna Images : The model trained on 15% lacuna images gives a validation CER of 7.83% and test CER of 11.11% on a mixture of lacuna and non-lacuna images. Table 1d shows the model's performance in predicting both the lacuna and non-lacuna characters in the dataset. In line with the model trained on 30% lacuna images, there is a decrease in the accuracy of predicting lacuna characters, while the accuracy for non-lacuna characters shows improvement. This suggests that reducing the proportion of lacuna images in the training data significantly limits the model's ability to learn about lacunae compared to training with 100% of lacuna images in the dataset. Another intriguing observation is the increase in the model's accuracy in correctly predicting clean non-lacuna characters from 77.82% (accuracy of a model trained solely on binarized non-lacuna images) to 80.35%. This suggests that the inclusion of additional lacuna images in the training set contributes to enhancing performance on non-lacuna characters in line images. It is possible that initially, the model might incorrectly predict a clean non-lacuna character. However, exposure to multiple copies of the same image (lacuna and non-lacuna versions) could have assisted the model in transcribing that character correctly.

5.3 Results from Our Log Probability and Attention Experiments

Our findings indicate that attention does not outperform the log probability baseline in identifying lacunae and other errors in the line images. Table 2 illustrates the efficacy of our logistic regression models trained on both log probability and attention entropy as features. We observe that attention entropy gets smaller feature importance coefficients compared to log probability in the classification of lacunae and transcription errors. The baseline models, using log probability as features, accurately predict whether a line image has any lacuna 53.25% of the time and errors 84.12% of the time. Figure 6 shows confusion matrices for all models trained to flag lacunae and errors in a line image. The model trained on log probability accurately identifies 396 non-lacuna images and 201 lacuna images. Similarly, it correctly flags 907 line images with transcription errors. Overall, log probability appears to be a robust metric to flag any lacunae and transcription errors in the line images.

In our analysis, we focus exclusively on the last layer and the first head of attention. However, it is essential to explore whether these findings remain consistent across all layers and heads, as well as with different attention metrics. Such an investigation could provide valuable insights into the robustness and generalizability of our results.

Table 2. Comparison of log probability and attention entropy features shows no difference in predictive accuracy.

Task	Feature Importance	Accuracy
Lacuna	Log probability: -0.247	53.25%
Lacuna	Log probability: -0.163 & Entropy: 0.001	53.25%
Other errors	Log probability: -2.683	84.12%
Other errors	Log probability: -2.688 & Entropy: 0.064	84.03%

5.4 Results from Training Using Leiden Conventions

We observe that the model generates square brackets (`[]`) in the transcription but consistently fails to enclose lacuna characters within those brackets. For instance, instead of `round a doll's h[o]use` it generates `round a [] doll's house`. It may be worthwhile to investigate if increasing the proportion of training data containing transcriptions formatted with the Leiden conventions, coupled with using an alternative loss function, could enable the model to learn and generate the appropriate bracketing. It may also be worth exploring the interaction of these conventions with the tokenization of the decoder model. Putting brackets around whole output tokens, for instance, could still indicate the presence of a lacuna in a word without modifying output tokenization.

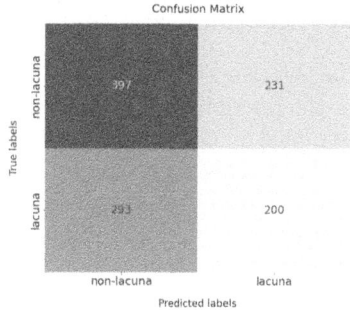
(a) Detecting lacunae with log probability and entropy features

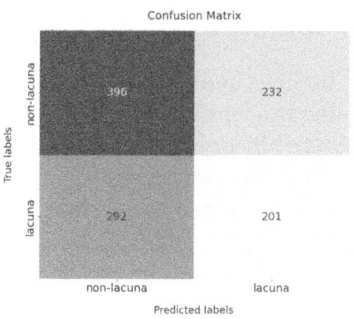
(b) Detecting lacunae with log probability feature

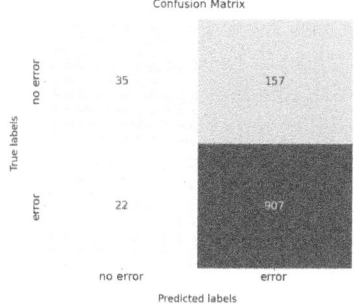
(c) Detecting errors with log probability and entropy features

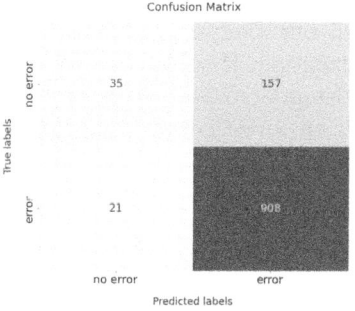
(d) Detecting errors with log probability feature

Fig. 6. Confusion Matrices for logistic regression models trained to flag lacuna and errors in line images with log probability and attention entropy as features.

6 Conclusion

Transformer models like TrOCR can successfully learn to restore lacunae in line images when trained on datasets featuring such gaps. Log probability can be leveraged to flag lacunae and transcription errors (e.g., due to non-standard writing or inking issues). Our controlled experimental setup demonstrates the promising potential of transformer-based models for restoring damaged historical documents. Moreover, by studying mechanistic properties such as log probability, it is possible to infer whether a specific document or image contains damages or inconsistencies like lacunae or errors without explicitly looking at those documents or images. Looking at more fine-grained attention features over a wider range of model representations and model architectures would be an interesting area of future work. Expanding our capacity for model interpretability could prove invaluable to scholars seeking to incorporate computational models into a paleographical workflow. Extending our methodology to real historical documents with lacunae will be the focus of our future work.

References

1. Arlitsch, K., Herbert, J.: Microfilm, paper, and OCR: issues in newspaper digitization. Utah Digit. Newspapers Program **33**(2), 59–67 (2004)
2. Assael, Y., et al.: Restoring and attributing ancient texts using deep neural networks. Nature **603**(7900), 280–283 (2022)
3. Baevski, A., Zhou, Y., Mohamed, A., Auli, M.: wav2vec 2.0: A framework for self-supervised learning of speech representations. In: Larochelle, H., Ranzato, M., Hadsell, R., Balcan, M., Lin, H. (eds.) Advances in Neural Information Processing Systems, vol. 33, pp. 12449–12460. Curran Associates, Inc. (2020)
4. Cowen-Breen, C., Brooks, C., Haubold, J., Graziosi, B.: Logion: machine learning for Greek philology. arXiv:2305.01099 [cs] (2023)
5. Devlin, J., Chang, M.W., Lee, K., Toutanova, K.: BERT: pre-training of deep bidirectional transformers for language understanding. In: Burstein, J., Doran, C., Solorio, T. (eds.) Proceedings of the 2019 Conference of the North American Chapter of the Association for Computational Linguistics: Human Language Technologies, Volume 1 (Long and Short Papers), pp. 4171–4186. Association for Computational Linguistics, Minneapolis, Minnesota (2019)
6. Dong, R., Smith, D.: Multi-input attention for unsupervised OCR correction. In: Gurevych, I., Miyao, Y. (eds.) Proceedings of the 56th Annual Meeting of the Association for Computational Linguistics (Volume 1: Long Papers), pp. 2363–2372. Association for Computational Linguistics, Melbourne, Australia (2018)
7. van Groningen, B.A.: De signis criticis in edendo adhibendis. Mnemosyne **59**(4), 362–365 (1932), https://www.jstor.org/stable/4426628
8. Hämäläinen, M., Hengchen, S.: From the paft to the fiiture: a fully automatic NMT and word embeddings method for OCR post-correction. In: Mitkov, R., Angelova, G. (eds.) Proceedings of the International Conference on Recent Advances in Natural Language Processing (RANLP 2019), pp. 431–436. INCOMA Ltd., Varna, Bulgaria (2019)
9. Koenecke, A., Choi, A.S.G., Mei, K., Schellmann, H., Sloane, M.: Careless whisper: speech-to-text hallucination harms. arXiv:2402.08021 [cs] (2024)
10. Li, M., et al.: TrOCR: transformer-based optical character recognition with pre-trained models. In: Proceedings of the AAAI Conference on Artificial Intelligence, vol. 37, no. 11, pp. 13094–13102 (2023)
11. Nockels, J., Gooding, P., Terras, M.: The implications of handwritten text recognition for accessing the past at scale. J. Documentation **80**(7), 148–167 (2024). https://doi.org/10.1108/JD-09-2023-0183
12. Radford, A., Wu, J., Child, R., Luan, D., Amodei, D., Sutskever, I.: Language models are unsupervised multitask learners. OpenAI blog **1**, 9 (2019)
13. Raffel, C., et al.: Exploring the limits of transfer learning with a unified text-to-text transformer. J. Mach. Learn. Res. **21**, 1–67 (2020)
14. Rijhwani, S., Anastasopoulos, A., Neubig, G.: OCR post correction for endangered language texts. In: Webber, B., Cohn, T., He, Y., Liu, Y. (eds.) Proceedings of the 2020 Conference on Empirical Methods in Natural Language Processing (EMNLP), pp. 5931–5942. Association for Computational Linguistics, Online (2020)

15. Rijhwani, S., Rosenblum, D., Anastasopoulos, A., Neubig, G.: Lexically aware semi-supervised learning for OCR post-correction. Trans. Assoc. Comput. Linguist. **9**, 1285–1302 (2021)
16. Vogler, N., Allen, J., Miller, M., Berg-Kirkpatrick, T.: Lacuna reconstruction: self-supervised pre-training for low-resource historical document transcription. In: Carpuat, M., de Marneffe, M.C., Meza Ruiz, I.V. (eds.) Findings of the Association for Computational Linguistics: NAACL 2022, pp. 206–216. Association for Computational Linguistics, Seattle, United States (Jul (2022)

NeuroPapyri: A Deep Attention Embedding Network for Handwritten Papyri Retrieval

Giuseppe De Gregorio[1](✉), Simon Perrin[2], Rodrigo C. G. Pena[1], Isabelle Marthot-Santaniello[1], and Harold Mouchère[2]

[1] University of Basel, Basel, Switzerland
{giuseppe.degregorio,rodrigo.cerqueiragonzalezpena,
i.marthot-santaniello}@unibas.ch

[2] Laboratoire des Sciences du Numérique de Nantes (LS2N),
Nantes Université, École Centrale Nantes, CNRS, LS2N,UMR 6004,
44000 Nantes, France
{simon.perrin,harold.mouchere}@univ-nantes.fr

Abstract. The intersection of computer vision and machine learning has emerged as a promising avenue for advancing historical research, facilitating a more profound exploration of our past. However, the application of machine learning approaches in historical palaeography is often met with criticism due to their perceived "black box" nature. In response to this challenge, we introduce NeuroPapyri, an innovative deep learning-based model specifically designed for the analysis of images containing ancient Greek papyri. To address concerns related to transparency and interpretability, the model incorporates an attention mechanism. This attention mechanism not only enhances the model's performance but also provides a visual representation of the image regions that significantly contribute to the decision-making process. Specifically calibrated for processing images of papyrus documents with lines of handwritten text, the model utilizes individual attention maps to inform the presence or absence of specific characters in the input image. This paper presents the NeuroPapyri model, including its architecture and training methodology. Results from the evaluation demonstrate NeuroPapyri's efficacy in document retrieval, showcasing its potential to advance the analysis of historical manuscripts.

Keywords: Deep Learning · Document Retrieval · Historical Document Analysis · Handwriting · Greek Papyri

1 Introduction

Greek papyri preserved thanks to the dryness of the Egyptian climate are an unrivalled source for historians and philologists, documenting many aspects of daily life but also preserving otherwise lost pieces of literature. However, papyri are most of the time fragmented, spread across collections around the world and lacking information on the context of their discovery. An essential work done by the papyrologists is to reconstruct fragmented papyri. Sometimes the content helps find joins when we can speculate what the text was (known literary text, recurrent formulaic wordings in documents). However, for some important cases like lost pieces of literature, finding similarities in visual layout and shape of handwriting is essential. Paleographers stumbled upon the difficulty of defining what they mean by "similarity". There is neither consensus on typologies of scripts nor on styles nor even on letter shapes. Literary papyri are not signed so it is only in exceptional cases that we can know their writers. Thus in this study, we used only elements upon which there is consensus: character identification based on transcriptions and line detection based on lines preserved on the same document.

The intersection of computer vision and machine learning (ML) presents a promising avenue for historical research, enabling a deeper exploration of our past. Nevertheless, ML approaches are often criticized for their perceived "black box" nature. To bridge the gap between paleographers and computer scientists, we introduce NeuroPapyri, an innovative deep learning network tailored to provide an embedding representation of an image of a papyrus written in ancient Greek. NeuroPapyri incorporates an attention layer with multiple attention heads, each specializing in focusing on distinct features within handwritten images. This attention mechanism aims to enhance transparency by revealing the specific regions of interest that influenced the model's decision-making process. The integration of attention mechanisms facilitates more meaningful exchanges between experts in palaeography and computer scientists, fostering collaborative advancements in the field of historical document analysis. In the following sections, Sect. 2 provides a brief overview of existing methodologies, Sect. 3 delves into the architecture of NeuroPapyri and the training methodology employed, Sect. 4 presents the datasets used for the experimentation. Section 5 presents the results obtained through rigorous testing. Section 6 is dedicated to an ablation study of the architecture. Finally, Sect. 7 and 8 focus on discussion and conclusion. Additional material, including the GitHub repository, is available at the link https://d-scribes.philhist.unibas.ch/en/case-studies/iliad-208/neuropapyri/.

2 Related Works

The retrieval of papyrus based on fragment analysis has been a relatively underexplored area within the scientific community. In recent years, a surge in interest was rather on the task of Writer Identification, leading to the proposition of several solutions. In their work, Christlein et al. [2] delve into the challenges associated with automatic writer identification and retrieval in Greek papyri. The study places significant emphasis on preprocessing techniques and feature sampling, highlighting the essential role of effective binarization in enhancing identification accuracy. Two unsupervised methods, one employing traditional features and the other leveraging self-supervised deep learning, are evaluated.

The research underscores the significance of writer identification in advancing studies in linguistics, history, and palaeography related to historical manuscripts. The approach involves local feature extraction using SIFT descriptors and self-supervised learning with a CNN, with binarization considered a critical step. Evaluation on the GRK-Papyri dataset [6] demonstrates improved accuracy with appropriate binarization. Peer et al. [7] propose a deep-learning-based approach for writer retrieval and identification on papyri fragments. The focus is on identifying fragments associated with specific writers and those corresponding to the same image. The authors introduce a novel neural network architecture that combines a residual backbone with a feature mixing stage to enhance retrieval performance. The methodology is evaluated on two benchmarks, PapyRow [3] and HisFragIR20 [11], achieving competitive results in writer retrieval and identification tasks. Rather than Writer Identification, Pirrone et al. [8] address the challenge of assembling fragmented papyri, offering a solution to assist papyrologists in reconstructing historical documents. The proposed method involves a deep siamese network architecture, named Papy-S-Net, designed for matching papyrus fragments. The model is trained and validated on a dataset of 500 papyrus fragments. The study explores various patch extraction approaches, demonstrating superior performance when trained on patches containing text. The method proves effective in a real-use case, achieving a 79% accuracy in matching fragments. Pirrone et al. [9] focus on a self-supervised deep metric learning approach for automatically associating ancient papyrus fragments, streamlining the challenging task of fragment reconstruction. The method, based on Deep Convolutional Siamese-Networks, demonstrates superior performance compared to domain adaptation, achieving a notable top-1 accuracy of 87% in a retrieval task on the HisFragIR20 dataset [11]. Emphasizing minimal human intervention, the study suggests that this approach holds promise for aiding papyrologists in fragment reconstruction without the need for extensive manual annotation or reliance on pre-trained models.

3 The Model

NeuroPapyri is crafted for the task of analyzing images of ancient Greek papyri, aiming to retrieve relevant information and provide an embedding representation. The architectural foundation is established upon an integration of Convolutional Neural Networks (CNN) and a purposefully engineered multi-head attention layer. Figure 1 displays the general architecture of the proposed model.

The model is specifically calibrated for processing images featuring lines of handwritten text. The architecture comprises an initial convolutional phase dedicated to feature extraction, followed by the incorporation of an attention block with multiple heads operating in parallel. As convolutional network, the ResNet-18 [5] has been used, as it is recognized for its ability to achieve high performance for a variety of problems in different domains. At the model's core lies the attention block, a crucial component facilitating simultaneous focus on diverse regions of interest within the image. The attention block features several heads,

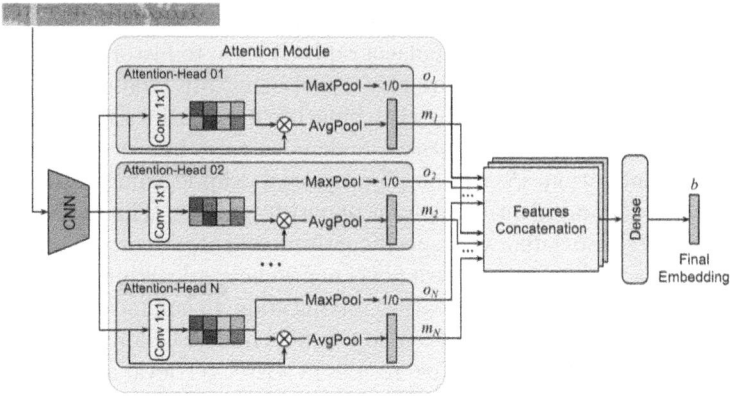

Fig. 1. Main architecture of the NeuroPapyri model.

each attention head functions autonomously, capturing specific information and thereby contributing to the formulation of a comprehensive representation. The underlying idea is to facilitate collaborative functionality among the attention heads, discerning unique strokes, curves, and details essential for accurate comprehension of the manuscript text. Each attention head produces both a feature map m_i and a binary output o_i. The computation of the feature map m_i involves a 1×1 convolutional layer designed to condense all features generated by the preceding convolutional phase into a singular attention map. A sigmoid activation function is used on the attention map to allow the values to lie between 0 and 1 and allow the map to pay equivalent levels of attention to different areas of the image, unlike what softmax would do. This map is then multiplicatively applied to the 1×1 convolution input, weighting each pixel of the feature map. The ensuing step involves the amalgamation of distinct feature maps derived from the attention module, which are subsequently subjected to processing by a feed-forward layer, culminating in the accomplishment of the final representation.

The generation of the network's attention maps can be trained with a weakly supervised approach, where the network autonomously refines its attention during training. As the network is trained, it will learn on its own to focus its attention on certain parts of the information to make its decision. However, the model introduces the possibility to enable the guidance of this learning phase through the binary outputs of the attention heads o_i. An attention loss is incorporated during the training phase, allowing for deliberate direction of attention map generation. This informative mechanism empowers the model to discern the presence or absence of specific characteristics within different training images.

3.1 Retrieval of Original Papyrus

The main goal of this work is to use the NeuroPapyri network to address the problem of document retrieval within historical papyri written in ancient Greek, recognizing that papyrus images are commonly organized as fragments rather than whole document images. This means that the image of a papyrus may represent a fragment of a larger document, so it raises the question of which original papyrus the fragmentary image corresponds to.

A segmented-based approach that takes into consideration images of the text lines of the document helps to focus on the paleographic features of the manuscript and therefore on the characteristics of handwriting, giving less weight to other features related to elements of the image which may suffer more the effect of different digitizing protocols (such as different brightness levels, different resolutions, and so on). This aspect is important to take into consideration, as typically the different images of papyrus fragments are preserved in different collections which have not harmonized their protocols and digitization processes. In this context, therefore, framing the problem as the recovery of a document based on the analysis of lines of handwritten text becomes a relevant consideration.

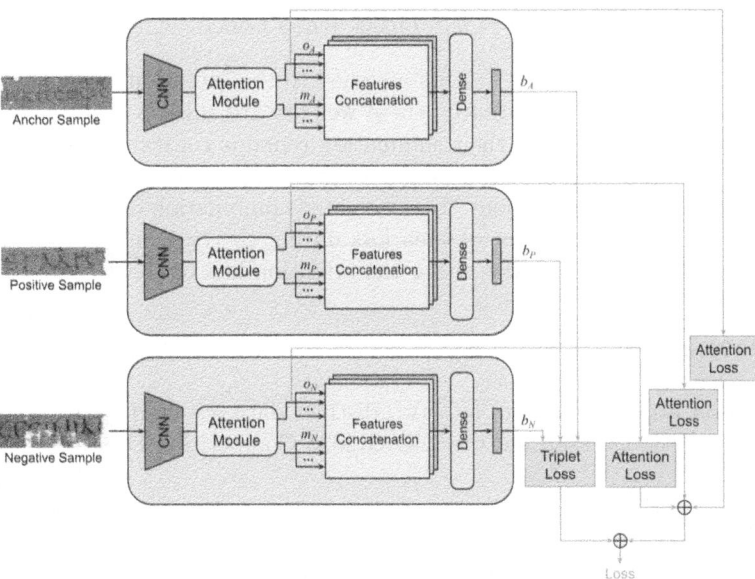

Fig. 2. NeuroPapyri can be trained using two different losses. *Attention loss* focuses on attention maps while *Triplet loss* aims to train the model to solve the main target problem.

To address the problem of document retrieval, the proposed model employs triplet learning to derive the final image representation. The architecture is implemented within a Siamese network framework [1], striving to acquire a discerning similarity measure between distinct images. Figure 2 displays the architecture in the Siamese configuration highlighting the application of the different losses. Through the application of a triplet loss, the model is trained to minimize the distance between row representations from the same document, while concurrently maximizing the distance between rows originating from different documents. Consequently, a query image can be associated to a specific document by evaluating its distance from the other images in the dataset. Notably, the proposed model offers a distinctive feature in its adaptability to two distinct training schemes. In the first scheme, the model autonomously determines attention map configurations, achieved by omitting consideration of attention loss during training. Alternatively, both attention loss $Loss_A$ and target loss $Loss_T$ can be harnessed concurrently to guide the generation of attention maps. For instance, the model can be directed to focus attention on images containing a specified set of characters, thereby affording a dynamic and tailored approach to training.

In order to afford the model the flexibility to assign distinct levels of importance to different losses, their combination is achieved through a weighted sum, as formulated by the following equation:

$$Loss = w_1 \cdot Loss_A + w_2 \cdot Loss_T \qquad (1)$$

Here, w_1 and w_2 denote the weights assigned to the attention loss and the target loss, respectively. Importantly, these weights are subject to the constraint $w_1 + w_2 = 1$, ensuring that their summation remains constant. The constraint ensures a normalized weighting scheme, facilitating a coherent and interpretable adjustment of loss contributions. This weighted combination of losses allows the model to adjust the influence of each loss during the training process and the model can effectively prioritize either the attention loss or the target loss.

4 Datasets

This section provides insights into two distinct datasets used for experimentation: a Synthetic Dataset and the ICDAR2023 Competition Dataset.

4.1 Synthetic Dataset

A synthetic dataset has been built for the preliminary evaluation stage. To create the synthetic dataset, the AL-PUBv2 dataset [12] has been used. AL-PUBv2 is a dataset containing images of handwritten ancient Greek characters collected by crowdsourcing. The images of this dataset have a very different quality due to preservation state, digitization methods or crowdsourcing operations. The synthetic dataset is built by pasting, side by side, 10 images from AL-PUBv2 resulting in 70×700 pixels images. The training set is made of 32768 images with transcription, while the testing set counts 1024 images. First strings of characters are formed randomly, and then images of each of those characters are taken randomly from AL-PUBv2. None of the character images used to build the test set are used in the training set. Figure 3 presents an example of a created image.

Fig. 3. Image from the synthetic dataset. Transcription: *ΦΟΟΠΣΜΛΑΓΝ*.

4.2 ICDAR2023 Competition Dataset

The "ICDAR 2023 Competition on Detection and Recognition of Greek Letters on Papyri" [10] has proposed a dataset containing 194 high-resolution images of papyri. All these papyri bear parts of Homer's *Iliad*. Each papyrus image is annotated by specialists at the character level. A bounding box is drawn around each character associated with its transcription and a quality tag indicating if the character is well preserved, incomplete but unambiguously recognizable or highly damaged.

Character distribution analysis is a fundamental aspect in the study of textual datasets, offering insights into the prevalence of individual letters and their linguistic significance. The character distribution of the ICDAR2023 dataset, as illustrated in Fig. 4a, showcases disparities in the representation of letters. Notably, certain letters exhibit lower occurrences, suggesting a skewed distribution influenced by the infrequency of their use.

While the primary focus of the competition does not coincide with the objective of document retrieval, the thoroughness of the dataset annotations renders it readily adaptable to the scope of our study. Given the scarcity of dedicated datasets tailored to the task of retrieving ancient papyrus documents, the ease of repurposing this dataset persuaded us to use it for our purpose.

5 Experimentation and Results

To assess the effectiveness of the NeuroPapyri model, a comprehensive series of experiments was undertaken, encompassing diverse datasets comprising papyrus images containing handwritten text in Ancient Greek. In this section, before presenting the results of experiments specifically addressing the retrieval of papyrus documents, we assess the network's performance on a preliminary and simpler task: character identification within text lines. This experiment enables us to ascertain whether the network, with its attention mechanism, effectively discerns the presence of specific characters in the image. Additionally, it offers insights into whether the decision is in any way correlated with the textual content of the images.

5.1 Character Identification

This evaluation scenario is designed to assess the presence of characters within a line of handwritten text. To tackle this challenge, the model can be trained by integrating a dedicated attention loss for character identification. In this setup, the model is settled with attention heads equal to the number of characters targeted for identification. The output o_i of a given attention head corresponds to 1 if the pertinent character is present in the image, or to 0 if the character is absent. Essentially, o_i represents a binary label, facilitating the modelling of this behaviour through the application of a binary cross-entropy cost function. Therefore, for the purposes of the experiment, the model is parameterized with 24 attention heads, each corresponding to a character of the Greek alphabet.

To evaluate the model, precision, recall and F1-score are used. A detection is considered positive when the network identifies a letter, regardless of how many such letters are in the image or where it puts its attention to identify the character. This evaluation serves as an investigation into the efficacy of attention loss, offering insights into its influence on the definition of attention maps.

Synthetic Dataset. Firstly, the model is trained on the synthetic dataset. Binary cross-entropy loss is used as attention loss. The Adam optimizer is used with a learning rate of $5 \cdot 10^{-5}$ and a batch size of 16 images is used.

Table 1 shows the results of character identification for the synthetic dataset. The F1-score of 82.14% shows that the model is able to identify characters in a synthetic papyrus line image using attention loss.

Table 1. Precision, Recall and F1-score for the model with synthetic and ICDAR2023 Competition dataset [10].

Dataset	Precision	Recall	F1-score
Synthetic dataset	78.15	86.55	82.14
ICDAR2023 dataset [10]	69.4	82.37	75.33

ICDAR2023 Competition Dataset. To evaluate the model on real papyri images, the ICDAR2023 competition dataset is employed. To align the dataset with the problem domain, we meticulously extracted 6,423 text line fragments, each accompanied by its corresponding content transcription from the original images. Labelling was performed in a manner that ensured consistency across lines originating from distinct images of the same document. Subsequently, an 80%/20% split was applied, designating 80% of the data for training and the remaining 20% for testing. Attention loss and Adam optimizer are still used with a learning rate of $1 \cdot 10^{-4}$ and mini-batches size of 4 images. Greyscale data augmentation is used with a 20% rate. The training lasted 18 epochs. Results from this experiment are present in Table 1. Remarkably, the training on real data yields results comparable to those achieved on the synthetic dataset. Figure 4b provides a visual representation of the identification rate for each character in the ICDAR2023 test set. We can note that some characters show lower identification rates than others. Characters such as Beta (B/β), Zeta (Z/ζ), Xi (Ξ/ξ), Psi (Ψ/ψ) and Phi (Φ/ϕ) show poor identification rates but they are also among the fewest represented within the dataset. Lambda (Λ/λ) and Gamma (Γ/γ) also show low identification rates despite their higher representation, potentially influenced by their shapes, which can be visually similar to other characters like Alpha (A/α) or Iota (I/ι). In general, it is possible to see how the identification rate is influenced by the level of representation of the character within the dataset. The most present characters show an identification rate which is on average higher than those which are not very frequent in the data set.

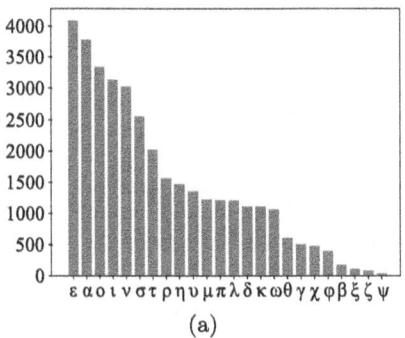
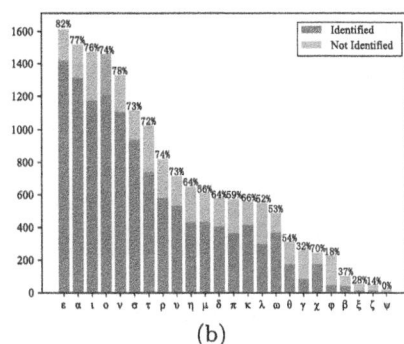

(a)   (b)

Fig. 4. a) Character distribution in the ICDAR 2023 competition training set; b) Identification rate for each letter in the ICDAR 2023 test set.

Figure 5 presents some attention maps generated by the network for images from the test set. Notably, the attention map for Eta (H/η) in the second row first column indicates that the network does focus its attention on the Eta. The attention is only on a small part of the letter, not on the whole letter. The attention maps of Sigma ($\Sigma/\sigma/\varsigma$) and Theta (Θ/θ) on the first column and of Alpha (A/α) on the last column show that the network focuses attention on two characters, suggesting that the network is able to identify multiple characters in an image. Additionally, the attention map dedicated to Mu (M/μ) in the last column shows an instance where the network fails to identify a character, as no attention is allocated to the corresponding region. This Mu was, however, partially damaged.

5.2 Document Retrieval

In this section, we present the results of our experiments conducted on the Document Retrieval problem. The dataset employed for this investigation corresponds to the one derived from the ICDAR 2023 competition. The used optimization technique is the Adam optimizer, with a learning rate set at $1 \cdot 10^{-4}$, and a batch size of 32 triplets of images. Cosine distance [4] was used to calculate the distance between image embeddings. The training process allowed the network to learn for a maximum of 50 epochs. While the system effectively generates an ordered list of documents, our emphasis lies in identifying the closest document

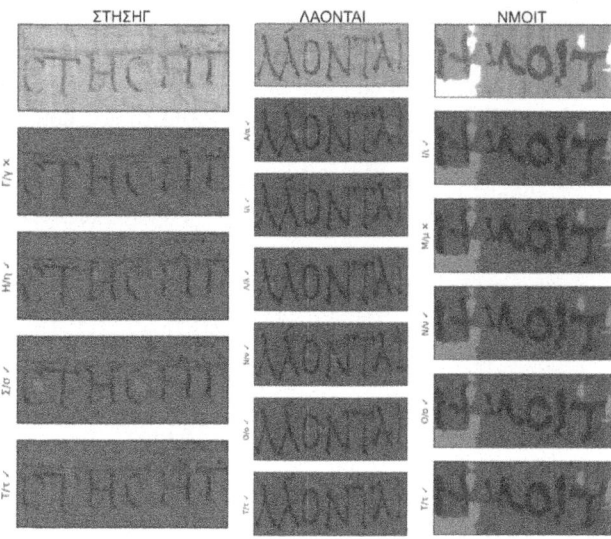

Fig. 5. Attention maps for some text lines of the ICDAR2023 dataset [10].

with meaningful relevance in our specific scenario. Each fragment is associated with a singular document, prompting us to assess the system's performance by considering only the closest document obtained. For this reason, the evaluation metrics reported focus on Top-1 Accuracy and F1@1.

As previously indicated, the network can be trained using two distinct methodologies, primarily differing in the incorporation of attention loss and the subsequent modelling of attention map generation. Initially, the network was trained without incorporating attention loss. Subsequently, the model was trained using attention loss, akin to the methodology employed in the character identification experiments. Specifically, each attention head was dedicated to a particular character, with attention maps being conditionally generated based on the presence or absence of the target characters in the image. Furthermore, the dual loss configuration was explored, as described by Eq. 1, where the combination of attention loss and target loss is regulated by the weights w_1 and w_2. Systematic variation of these weights was conducted to assess the influence of each loss on the overall learning trajectory. Experimental findings highlighted a notable divergence in the attention loss when the weight w_1 exceeded 0.50, adversely impacting the system's performance on the validation set. The graphical representation (see Fig. 6) illustrates the attention loss trend during training with 24 attention heads as the weights w_1 vary. Notably, the loss on the validation set tends to diverge with increasing values of w_1. Table 2 presents the quantitative results. The performance on the test set demonstrates improvement for lower values of the attention loss weight (w_1), underscoring the importance of attention loss weighting in shaping the system's overall efficacy.

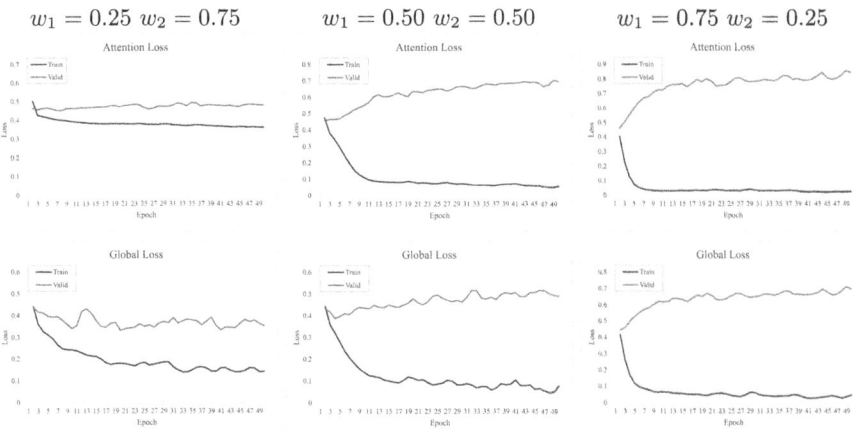

Fig. 6. Evolution of loss during training for different loss weights.

Table 2. Top-1 Accuracy and F1@1 score of the model as the combination weights of the loss w_1 and w_2 vary

	$w_1 = 0.25$ $w_2 = 0.75$	$w_1 = 0.50$ $w_2 = 0.50$	$w_1 = 0.75$ $w_2 = 0.25$
$Top1$	**92.82**	91.47	89.71
$F1@1$	**89.04**	88.07	84.21

The determination of the number of attention heads is not hardcoded in the model architecture. Consequently, the model was trained considering diverse numbers of attention heads, ranging up to a maximum of 24, corresponding to the number of letters in the Greek alphabet. Notably, configuring the number of attention heads in the double loss paradigm involves nuanced decisions. A critical consideration in this regard is the selection of distinct attention heads and their association with specific characters. In our experiments, each attention head is intricately linked to the presence of a particular character in the image of the text line. To delineate the allocation of attention heads, we considered the frequency distribution of letters within the training set. Subsequently,

Table 3. Top-1 Accuracy and F1@1 score of the trained model with and without attention loss and when varying the number of attention heads

		N-Head = 1	N-Head = 3	N-Head = 5	N-Head = 10	N-Head = 24
With Attention-Loss	$Top1$	95.77	94.33	93.14	92.42	91.39
	$F1@1$	92.82	90.66	89.12	86.91	88.12
Without Attention-Loss	$Top1$	**96.57**	94.18	94.66	94.90	92.82
	$F1@1$	**94.00**	93.64	91.55	91.97	89.04

attention heads were selected based on the characters that exhibited the highest frequency of appearance. For instance, in the scenario of a model equipped with three attention heads, the selection process involved considering the three characters that appeared most frequently in the training set (ε, α, o), and so forth. Table 3 provides a comprehensive depiction of the model's performance, showcasing results for both models trained with and without attention loss across varying numbers of attention heads.

Comparative Investigation. To finish this section, we present the results of a comparative investigation conducted between our proposed model and two contemporary methodologies. Specifically, we conducted training procedures on the model introduced by Peer et al. [7] and the model presented by Pirrone et al. [9] on the ICDAR2023 dataset, which served as the basis for our evaluation. The tabulated results of this comparative analysis are outlined in Table 4. Notably, our model exhibited superior performance in contrast to the models under consideration. The methodology proposed by Peer et al. shares similarities with our approach, as it focuses on the analysis of text lines. Conversely, Pirrone et al.'s model operates on random crops extracted from documents of square dimensions. This methodological divergence raises concerns regarding the consistency of the training data, as it does not ensure the exclusivity of handwritten text within the images, thereby complicating the comparison process.

Nonetheless, it is important to acknowledge the challenges associated with conducting a comprehensive comparative assessment. Limitations arise from the scarcity of bench-marking methodologies and the inherent discrepancies in dataset composition. Furthermore, variations in methodological approaches across different works contribute to the intricacy of bench-marking procedures.

Table 4. Comparison with State-of-the-Art. Top-1 Accuracy and F1@1 score.

Method	Top-1	F1@1
Peer et al. [7]	71.16	70.65
Pirrone et al. [9]	88.89	-
NeuroPapyri	**96.57**	**94.00**

6 Ablative Study

In this section we present an ablative study with the aim to provide valuable insights into the influence of various components of our model on overall performance, allowing us to evaluate whether the different components of the system can contribute positively or negatively to solving the problem. Specifically, we focus on two key characteristics: the adoption of a Siamese architecture and the impact of the proposed attention module.

To establish a baseline for comparison, we trained the model using only the convolutional stage. The outcomes of this experiment serve as a benchmark for evaluating our model's performance and assessing the efficacy of design choices made during the ideation phase. Subsequently, we trained our model based on a Siamese architecture, excluding the proposed attention module. These results are then compared with those obtained by training the complete model, integrating both Siamese architecture and the attention module. Table 5 reports the results of these different model configurations, facilitating conclusions drawn from the ablative study. Looking at the table, it becomes evident that the adoption of the Siamese architecture alone leads to a significant enhancement in performance, manifesting as a noteworthy 13-point increase in the F1@1 index. Moreover, the introduction of the attention block serves to further refine performance indices, highlighting its significant contribution to the model.

Table 5. Top-1 Accuracy and F1@1 score for the ablation study

	Top-1	F1@1
ResNet-18	90.19	77.10
ResNet-18 + Siamese	93.94	90.42
ResNet-18 + Siamese + Attention	**96.57**	**94.00**

7 Discussion

This section aims to discuss the implications and limitations arising from the findings and methodology presented in this study.

The incorporation of a multi-head attention mechanism in NeuroPapyri provides insight into an interpretability study of network behaviour, a critical aspect often lacking in machine learning models. Figure 7 shows the attention maps when the network is trained to solve the Document Retrieval problem, where the parts of the image where attention is concentrated are highlighted in red. Observing the attention maps reveals a consistent trend of activation towards the text, indicating the significance of handwriting amidst background elements. Notably, the attention maps exhibit a broader scope of interest beyond solely the target character, often encompassing a segment of writing larger than the character itself. Furthermore, it can be seen from Table 2 that the best performance is achieved with only one head of attention. This may suggest that, to optimize performance, the network prefers not to focus on individual characters, but to analyze the writing as a whole.

The attention maps provide insights into the regions of interest that influence decision-making. The analysis of those maps can facilitate the collaboration between experts in palaeography and computer science, promoting a deeper understanding of the ancient Greek manuscript writing styles.

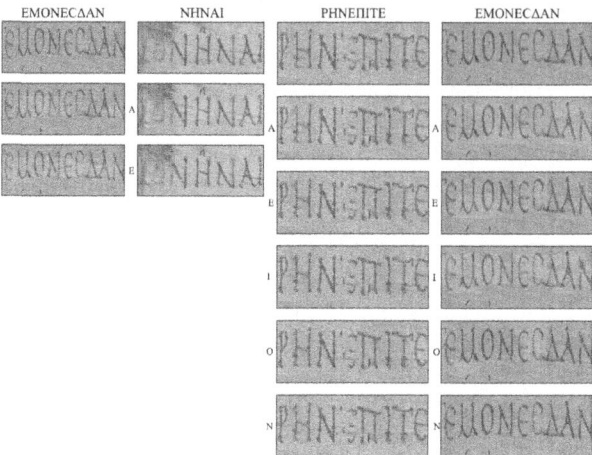

Fig. 7. Attention maps for some text lines. The first two images are the results of the system configured with two attention heads (the first for the letter A and the second for the E), while the last two are of a system with five attention heads (for the letters A,E,I,O,N).

While NeuroPapyri demonstrates promising capabilities, certain limitations must be acknowledged. The model's performance heavily relies on the quality and diversity of training data. Specifically, when analyzing the outcomes of the preliminary test of character identification, a notable trend emerges: the model's effectiveness in identifying individual characters correlates with the frequency of appearance of each character within the training set. Characters that occur more frequently in the training data tend to exhibit superior identification performance compared to those with limited representation. This observed pattern in the performance of character identification could potentially influence the distinctive contributions made by each attention head to the broader document retrieval problem. Examining the results of the document retrieval, it becomes evident that employing 24 attention heads may not be conducive to optimal performance. Furthermore, the distribution of attention does not fully align with the expectation that importance is given to each individual character. Rather, the network prefers a more general view of the text, trying only to avoid information belonging to the background of the papyrus.

8 Conclusion

In this work, we have presented a novel deep learning-based model, NeuroPapyri, designed for the analysis of images of ancient Greek papyri. The primary objective of the model is to provide an embedded representation of an image containing handwritten text in ancient Greek through the use of an attention mechanism capable of improving the performance of the model and at the same time providing a visual representation of the areas of the image that contributed most to the decision process.

The model was trained to address the problem of document retrieval. Using a Siamese network framework, the model reached promising results achieving a Top-1 accuracy of 96.57% on the dataset used for the ICDAR 2023 "Competition on Detection and Recognition of Greek Letters on Papyri". The innovative inclusion of an attention block further refined the model's performance, allowing it to learn details and relationships within handwritten text. The attention block generates visible maps that could allow scholars to evaluate which areas of the image contributed most to the decision. Through systematic experimentation, we explored the impact of attention block on system performance, recording an actual improvement in results. The model's effectiveness in character identification was demonstrated through experiments on two distinct datasets. By employing 24 attention heads, each corresponding to a letter in the Greek alphabet, NeuroPapyri achieved promising results showcasing the model's ability to focus attention on identified letters within the images.

The development of NeuroPapyri opens avenues for future research directions. Further refinement of attention mechanisms and exploration of advanced architectures may improve the model's robustness and adaptability to diverse historical datasets. The impact of losses on the learning trajectory deserves a more in-depth analysis to understand the best way to combine the different losses and the real contribution each loss makes to the learning process. NeuroPapyri's adaptability can be further tested by evaluating the network on different problems, such as Writer Identification or dating of papyri.

In conclusion, NeuroPapyri is a promising deep-learning framework for analysing handwritten papyri in ancient Greek. By combining cutting-edge technologies in computer vision and machine learning, this model contributes to the evolving landscape of historical document analysis, fostering collaboration between experts in palaeography and computer scientists. The transparency provided by the attention mechanism promotes meaningful exchanges and opens avenues for further advancements in the interdisciplinary field of historical research.

References

1. Chicco, D.: Siamese neural networks: an overview. Artif. Neural Networks, 73–94 (2021). https://doi.org/10.1007/978-1-0716-0826-5_3
2. Christlein, V., Marthot-Santaniello, I., Mayr, M., Nicolaou, A., Seuret, M.: Writer retrieval and writer identification in Greek papyri. In: Carmona-Duarte, C., Diaz, M., Ferrer, M.A., Morales, A. (eds.) IGS 2022. LNCS, vol. 13424, pp. 76–89. Springer, Cham (2022). https://doi.org/10.1007/978-3-031-19745-1_6
3. Cilia, N.D., De Stefano, C., Fontanella, F., Marthot-Santaniello, I., Scotto di Freca, A.: PapyRow: a dataset of row images from ancient Greek papyri for writers identification. In: Del Bimbo, A., et al. (eds.) ICPR 2021. LNCS, vol. 12667, pp. 223–234. Springer, Cham (2021). https://doi.org/10.1007/978-3-030-68787-8_16
4. Gomaa, W.H., Fahmy, A.A., et al.: A survey of text similarity approaches. Int. J. Comput. Appl. **68**(13), 13–18 (2013). https://doi.org/10.5120/11638-7118

5. He, K., Zhang, X., Ren, S., Sun, J.: Deep residual learning for image recognition. In: Proceedings of the IEEE Conference on Computer Vision and Pattern Recognition, pp. 770–778 (2016). https://doi.org/10.1109/CVPR.2016.90
6. Mohammed, H., Marthot-Santaniello, I., Märgner, V.: GRK-papyri: a dataset of Greek handwriting on papyri for the task of writer identification. In: 2019 International Conference on Document Analysis and Recognition (ICDAR), pp. 726–731 (2019). https://doi.org/10.1109/ICDAR.2019.00121
7. Peer, M., Sablatnig, R.: Feature mixing for writer retrieval and identification on papyri fragments. In: Proceedings of the 7th International Workshop on Historical Document Imaging and Processing. HIP '23, pp. 31–36 (2023). https://doi.org/10.1145/3604951.3605515
8. Pirrone, A., Aimar, M.B., Journet, N.: Papy-s-net: a Siamese network to match papyrus fragments. In: Proceedings of the 5th International Workshop on Historical Document Imaging and Processing. HIP '19, pp. 78–83. Association for Computing Machinery, New York, NY, USA (2019). https://doi.org/10.1145/3352631.3352646
9. Pirrone, A., Beurton-Aimar, M., Journet, N.: Self-supervised deep metric learning for ancient papyrus fragments retrieval. Int. J. Doc. Anal. Recogn. (IJDAR), 1–16 (2021). https://doi.org/10.1007/s10032-021-00369-1
10. Seuret, M., et al.: ICDAR 2023 competition on detection and recognition of Greek letters on papyri. In: Fink, G.A., Jain, R., Kise, K., Zanibbi, R. (eds.) ICDAR 2023. LNCS, vol. 14188, pp. 498–507. Springer, Cham (2023). https://doi.org/10.1007/978-3-031-41679-8_29
11. Seuret, M., Nicolaou, A., Stutzmann, D., Maier, A., Christlein, V.: ICFHR 2020 competition on image retrieval for historical handwritten fragments. In: 2020 17th International conference on frontiers in handwriting recognition (ICFHR), pp. 216–221. IEEE (2020). https://doi.org/10.1109/ICFHR2020.2020.00048
12. Swindall, M.I., et al.: Exploring learning approaches for ancient Greek character recognition with citizen science data. In: 2021 IEEE 17th International Conference on eScience (eScience), pp. 128–137 (2021). https://doi.org/10.1109/eScience51609.2021.00023

MONSTERMASH: Multidirectional, Overlapping, Nested, Spiral Text Extraction for Recognition Models of Arabic-Script Handwriting

Danlu Chen[1(✉)], Jacob Murel[2], Taimoor Shahid[3], Xiang Zhang[1], Jonathan Parkes Allen[3], Taylor Berg-Kirkpatrick[1], and David A. Smith[2]

[1] University of California, San Diego, CA, USA
dac013@ucsd.edu
[2] Northeastern University, Boston, MA, USA
[3] University of Maryland, College Park, MD, USA

Abstract. Most current models for handwritten text recognition transcribe individual lines and thus depend on accurate line extraction from page images. This line extraction task is particularly challenging for Arabic-script manuscripts, which exhibit a high proportion of curved lines, word baselines that vary within the line, and varying line orientation on the page. We present a new corpus for studying Arabic-script line extraction in the presence of these phenomena and evaluate different model architectures using several pixel-level, object-level, and extrinsic recognition metrics. Training all models on the same data, we find that the CNN-based Kraken model slightly outperforms the transformer-based TESTR model on recognition character accuracy and some object-level metrics, even though it lags behind on pixel-level metrics.

Keywords: Islamicate manuscripts · Text line extraction · HTR

1 Introduction

Methods for handwritten text recognition (HTR) have an array of applications for advancing research on historical manuscripts—from enabling full-text search of library collections, to conducting large-scale linguistic analyses of digitized documents. HTR has made significant progress since adopting line-level transcription approaches such as connectionist temporal classification (CTC). Just as earlier word-spotting approaches required segmenting a page image into word patches, line-level models require that the lines of text be extracted from a page to produce both training and test data. Although whole-page encoder-decoder models have been proposed, their successful deployment still requires, at the current state of the art, a set of phased training steps involving individual lines.

While there are many annotated datasets and shared-task evaluations to choose from for the task of transcribing images of single lines, annotated datasets and published evaluations of line extraction are less common. In any case, line extraction for handwritten text in Latin, Hebrew, Chinese, and other scripts can achieve a high degree of accuracy. The effectiveness of line extraction for Arabic-script manuscripts—in Arabic, Persian, Ottoman Turkish, Urdu, and other languages—seems anecdotally to be lower and less well explored.

This paper, therefore, seeks to fill this gap and provide a systematic analysis of the effectiveness of different line-extraction approaches for Arabic-script manuscripts. In the literature, these models have been evaluated on scripts such as Latin and Hebrew with various levels of success. Our contributions are both *observational*, resulting in a compilation of existing and new annotated data, and *experimental*, evaluating the effectiveness of pixel-labeling and object-detection approaches and convolutional and transformer neural architectures.

To perform our evaluations, we compile existing annotated data for Arabic-script manuscripts and also produce a newly annotated test set. Unlike some existing annotated datasets, which may omit marginalia and confine themselves to main-text horizontal lines, our new test set marks all lines of written text on each page and balances samples from simpler and more complex page layouts.

We evaluate three line-extraction approaches, training and testing them on the same data. First, `kraken` [10] labels pixels on a page as belonging to a baseline and then, in a postprocessing step, heuristically infers a polygon that bounds the characters on that baseline. Second, `doc-ufcn` [5] labels pixels as belonging to a line region. Third, `TESTR` directly infers a polygon bounding a line region.

We find that the pixel-based `kraken` performs best according to the object-level AP@0.75 metric, while the detection-based method `TESTR` performs best according to pixel-level metrics. While `TESTR` is slower than our other models, it is also the most accurate on the most difficult line extraction examples: nonlinear text with extreme curvature. In terms of downstream recognition accuracy, `TESTR` markedly outperforms `doc-ufcn`, but is marginally worse than `kraken`. Our results show that each system has its own strengths and weaknesses. We hope our provided datasets encourage further work on Arabic-script HTR for complex document layouts.

2 Related Work

2.1 Text Line Extraction

As mentioned, a critical initial step in most HTR pipelines is text line extraction. In the literature, there are three major approaches to line extraction: baseline-based [10], region-based [6], and detection-based [23] methods.

Baseline-based methods identify which pixels in the input image correspond to a baseline of a text line in the image. After the baselines have been identified, a post-processing step infers the surrounding polygon for each text line. Region-based methods also make pixel-level predictions, but instead of identifying baselines, they identify all pixels within a text line region. Detection-based methods, on the other hand, instead directly predict the vertices of polygons surrounding each text line.

The former two methods show suitable performance for horizontal lines, which has been the primary focus of past research [1,7,21]. Natural scene text extraction has similarly focused on linear text [12,18]. Non-linear lines, however, pose a significant challenge for these systems.

Previous approaches to detecting non-linear text lines have involved character-level bounding boxes [3] and semantic segmentation tools [13,15]. While these approaches show promise, they require computationally expensive pre- and post-processing steps. By contrast, transformer-based text-detection promises an effective and less expensive approach for non-linear text line extraction [9,20,23]

2.2 Arabic Script HTR

As with much language-focused research, HTR experiments have traditionally focused on Latinate texts-e.g. English, Spanish, French, etc. In the past few years, however, researchers have turned to non-Latinate documents. Arabic-script HTR poses several unique challenges due to its connecting script, use of diacritical marks, and changing character forms [11,14].

Existing Arabic-script HTR datasets include IFN-ENIT databse[1], AHDB [2], and KHATT [16]. Each of these, however, consists of images of text regions demarcated at the word or paragraph level sourced from modern writers for HTR research. As such, text lines are straight and undisturbed by peripheral text regions occupying image space.

Other studies that use datasets of historical Arabic-script documents do show promising work for Arabic-script text line extraction [1,21]. Unfortunately, these studies similarly focus on documents with exclusively horizontal lines. Moreover, neither publicly release their respective datasets for use, making it difficult to replicate and build upon their work.

We know of only two other study that exclusively addresses non-linear text line extraction for handwritten Arabic-script. [22] proposes partial contouring and projection to recognize curved text lines as horizontal segments and concatenate these segments for transcription. This approach evidences useful results. Given its dated-ness, however, it does not take advantage of the significant developments in test line extraction from the past decades. [4] also collected a curved text line dataset but only pixel-wise labels are available.

3 Dataset Description

3.1 Dataset Overview

The different subsets of ITI-bench are summarized in Table 1. There are two subsets for training and the other two for testing. We estimated the quality of the data by sampling 20 pages from each dataset and manually counted the number of missing lines or incorrectly labelled lines. We calculated the missing

---

[1] http://www.ifnenit.com/.

ratio for main body text (main) and marginalia separately, which is reported in the last two column in Table 1. 7.5% of the main body text is missing in the `arabic_ms` set.

Table 1. Overview of dataset statistics. **bsln.** standards for number of pages have valid baseline annotation and **poly.** for number of pages that have valid polygon annotation.

subset	usage	pages		HT	trans.	lines	drop rate	
		bsln.	poly.				body	margin
aocp_print	train	2,970	2,718	No	Yes	67,750	2.7%	17.3%
arabic_ms	train	873	873	Yes	Yes	16,635	7.5%	15.7%
aocp_ms_eval	test	113	112	Yes	Yes	2,025	0.0%	22.4%
iti-complex	test	52	52	Yes	No	2,536	0.0%	0.8%

1. `aocp_print`, annotated printed documents, total about 2,970 pages. Only 2,718 of them can be converted into polygons automatically.
2. `arabic_ms`, 873 annotated pages.
3. `aocp_ms_eval`, 113 annotated pages with transcription, https://github.com/OpenITI/aocp_ms_eval
4. `iti-complex`, a complex layout selection of Islamicate manuscripts.

3.2 Annotation Protocol

The `iti-complex` subset is a newly annotated set of line extractions that we provide. Our goal was to provide a more challenging benchmark for line extraction in this domain that more accurately represents the diversity of page layouts and line orientations. Thus, `iti-complex` consists of a set of 52 single page images selected from a wide range of manuscripts held within a diversity of digital repositories large and small. We aimed to represent multiple vectors of that diversity as much as possible within the constraints of the data set, placing an emphasis on manuscript pages with high degrees of layout complexity and line and word-form variability (that is, complex ligatures, full vocalization, tall ascending and descending letter forms, and so forth). We also captured a high degree of chronological, regional, and linguistic diversity, sampling not only Arabic and Persian but also Ottoman Turkish, Urdu, and Uyghur texts, along with multi-script texts including Coptic, Sanskrit, and Latin scripts. Generic diversity was somewhat constrained by our privileging high complexity layouts, although there too we included a wide range of texts. Finally, we also sought to represent the diversity of digitized exemplars, which range from high quality scans typical of recent digitization efforts to much lower quality, lower resolution and often grey scale or even black-and-white scans that are more typical of early digitization efforts or represent the digitization of microfilm.

Fig. 1. Example pages from iti-complex (image id 3, 6, 20, 23, 37, 49). There are not only vertical or diagonal aligned text lines, but also spiral, curved text lines.

Each page in this set is annotated with a polygon of up to 16 vertices using Labelme [19] by our experts. After selecting a page for use in the dataset, we saved a JPEG or PNG file, added the metadata relevant to the manuscript in question along with brief descriptions of layout features, and opened the image in Labelme. While we did practice some export and editing of previously annotated images (generated using the transcription platform eScriptorium, and then geometrically simplified), most of the material in our data set was annotated using previously un-annotated images taken directly from publicly open online repositories. We then drew the polygons within the application, maintaining a maximum of sixteen points, and including all letter-forms and orthographic devices as well as 'punctuation' marks. We treated words within a continuous line as units for polygon generation, which entailed in some cases breaking up continuous semantic units which however were not spatially continuous. In general this approach allowed us to capture the entirety of the letter-forms while minimizing neutral space and overlap with lines above and below (and on occasion intersecting in some way from the left or right). On the other hand, we treated lines in two columns that were a semantic unit as one continuous line. It is common in the Islamicate poetic tradition, which frequently employs two-line poetry units, to inscribe the two lines in separate columns. However, these are not true columns and are ignored while reading the text as the reading order involves reading the first line from the right column and jumping to the next column (left), and then jumping back to the first column for the next line. As such, for these cases, we treated the two lines in two columns as one. On occasion we were forced to make educated guesses about the relationship of a given dot to two lines, a not uncommon occurrence in Arabic script, but such instances were rare. Other edge cases included lines in which poor ink quality or damage to the manuscript over time made the precise decipherment of a line and its letter-forms difficult if not impossible. Finally, we had to make determinations about the occasional use of superscript letters used for in-text annotation and other purposes, choosing to include them as part of the line which they modified.

Figure 1 shows several examples from the ITI-complex subset.

4 Methods

There are three major kinds of line extraction methods: *baseline-based*, *region-based*, and *detection-based*. Both *baseline-based* and *region-based* methods produce pixel-wise predictions: *baseline-based* approaches identify pixels that represent the baseline of each text line, while *region-based* methods identify pixels that makeup the entire text line region. (In the computer vision community, models that predict dense pixel-wise outputs are also called segmentation-based methods.) In contrast, *detection-based* methods predict the vertices of polygons that contain each individual text line. In order to compare these paradigms, we train and test three representative line extraction models: kraken (baseline-based) [10], doc-ufcn (region-based) [6] and TESTR (detection-based) [23] (Table 2).

Table 2. A summary of the three kinds of line extraction methods. The **overlap** column indicates whether the method can represent overlapping line regions, while the **post-process** column indicates whether the backbone model's output requires post-processing in order to specify lines. Finally, the **train iter.** column shows the number of training iterations needed for each model.

	type	output	overlap	post-process	backbone	train iter.
kraken	baseline	binarized pixels	No	Yes	convnet	<50,000
doc-ufcn	region	binarized pixels	No	Yes	convnet	10,000
TESTR	detection	vertices of polygon	Yes	No	transformer	200,000

4.1 Kraken and DOC-UFCN

Kraken is an open-source OCR system that specializes in historical and non-Latin scripts. Both Kraken's line detection module and DOC-UFCN are U-NET-based models which consist of sequences of convolutional layers followed by transposed convolutional layers. A U-Net-like model generates 2D pixel-wise predictions that maintain the same dimensions as the input. Each pixel receives a numerical prediction ranging from 0 to 1. Kraken specifically predicts baselines, whereas DOC-UFCN classifies all pixels within the text line polygon as positive. For post-processing, Kraken employs heuristic methods to expand the baseline into a bounding box, while DOC-UFCN identifies connected components in the 2D prediction heatmap to extract polygons.

4.2 TESTR

TESTR is a transformer-based text detection model and predicts vertices of polygons directly. As the predictions does not restricted to be non-overlapping, or sometimes they can be identical, therefore we need to remove the redundant polygons. We perform non-maximum suppression (NMS) to reduce overlapping polygons and keep only the polygons with highest scores.

5 Experiments

For Kraken, we inference on the model checkpoints released by [17]. For the rest two models, we trained from scratch using similar training procedure.

5.1 Pre-processing

DOC-UFCN requires shrunken polygon to achieve the best performance. We set the shrink ratio to 0.1. We resize the image so that the width or height is no longer than 768 px for training data.

TESTR requires polygons with no more than 16 vertices. There are usually more than 40 vertices for each polygon for all subsets expect the **iti-complex** subset. We use the Douglas-Peucker algorithm [8] to simply the polygon and then compute the convex hull of the polygon.

5.2 Training

For DOC-UFCN and TESTR training, we followed Kraken's training recipe for fair comparison. We first train on `AOCP_print` (phase I) and then fine-tuning on `arabic_ms` (phase II).

DOC-UFCN. We trained from scratch with a learning rate 5^{-5} and batch size 32. We trained for 10,000 iterations.

TESTR. In Phase I, we trained from scratch with a learning rate 5^{-4} and batch size 8. We trained for 200,000 iterations. In Phase II, we trained on the last checkpoints from Phase I with a learning rate 5^{-5} and batch size 8. We trained for 200,000 iterations.

5.3 Inference

Kraken. If any side of the image is greater than 3000px, we need to resize it first to feed to Kraken. We simply using the default setting of Kraken for inference.

DOC-UFCN. We use the default setting of the model for inference.

TESTR. During influence, we set the confidence threshold to 0.05, i.e. we ignore predictions whose confidence scores are lower than 0.05. We perform non-maximum suppression (NMS) to reduce overlapping polygons. The IoU threshold for running NMS is 0.3.

5.4 Evaluation Metrics

Following [7], we use both *pixel-level* and *object-level* metrics for intrinsic evaluation of extracted lines, while we use *downstream OCR performance* as an extrinsic evaluation. In contrast with [7], however, we use macro-averaging across lines in all pixel-based metrics (i.e. those that compare areas of regions) to avoid issues arising from double counting overlapping line regions.[2]

Pixel-Level Metrics. These metrics are also used widely in general semantic segmentation tasks in the computer vision community. They evaluate the prediction performance aggregating across individual pixels in predicted and ground truth regions. However, these metrics sometimes fail to reflect whether a complete line region has been detected accurately.

1. **(macro) IoU.** For each page, we first compute the union of all the predicted line regions, `predict_union`, and similarly, compute the union of all ground truth line regions, `label_union`. Finally, the intersection over union (IoU) metric is given by calculating the ratio of the areas of the intersection and union of `predict_union` and `label_union`.
2. **(macro) recall/precision/F_1.** Similarly, recall is given by computing the ratio of the area of the intersection of `predict_union` and `label_union` with the area of `label_union`, while precision is calculated in a similar fashion but using the area of `predict_union` as the denominator. F_1 is the geometric mean of precision and recall.

---

[2] Alternatively, it would also be possible to align predict and groundtruth regions first, and then calculate micro-averaged pixel-based metrics. However, this cause the pixel-based metrics somewhat redundant with the separate object-based metrics we calculate.

Object-Level Metrics. These metrics are commonly used to evaluate object detection tasks in computer vision. We first greedily map predicted line regions to ground truth regions to maximize IoU, and then set an IoU threshold to determine which predicted regions can be considered correct. After determining the correct and incorrect predicted line regions based on the chosen threshold, we compute object-level precision, recall, and F_1. By varying the IoU threshold, we can compute metrics with variable tolerance in determining 'correctness'. Further, by varying the underlying threshold of model score that determines the confidence level at which the model makes a prediction, we can measure the tradeoff between precision and recall.

1. **AP (Average Precision).** This commonly used metric in the objective detection literature is derived by plotting the object-level precision-recall curve while sweeping the model confidence threshold across its full range and then computing the area under this curve. AP@.5 refers to the area under the precision-recall curve when the intersection over union (IoU) threshold is set to 0.5.

OCR Performance. Finally, as an extrinsic metric of line extraction performance, we pass the extractions of the predicted line regions to a downstream OCR engine and measure OCR performance against a groundtuth transcription. Specifically, we calculate the character accuracy rate (CAR) for each line extraction model using the recognition model from [17] as an OCR engine.

Table 3. Result on `aocp_ms_eval`. **Manual** reports the CAR when feeding the recognition system with manually labelled gold bounding boxes as a reference.

	pixel-level				object-level		OCR
	IOU	P	R	F_1	AP@.5	AP@.75	CAR
kraken	83.04	98.39	80.95	87.77	68.27	**56.49**	**65.08**
doc-ufcn	62.93	94.29	65.78	76.78	68.55	13.23	47.42
TESTR	91.21	87.00	98.95	92.60	**76.74**	32.09	63.83
manual							77.94

5.5 Results

The results on three subsets of the ITI dataset. Table 3 shows performance on `aocp_ms_eval`. As summarized in Table 3, about 25% of the annotation for marginalia is missing and therefore, the performance of TESTR–the method can predict fairly good on marginalia is penalized, as shown in Fig. 2. Note that TESTR predict the polygon directly, therefore, the IOU is calculated by the IOU of the union of predicted polygons and the union of labels.

We run the recognition model from [17] on the exacted lines from different models and report the character accuracy rate (CAR). We take the average of CAR from each page. Note that some of the labelled data does not have any manual transcription, and we just skip it (Table 4).

Table 4. Result on `iti-complex`.

	pixel-level				object-level	
	IOU	Precision	Recall	F_1	AP@.5	AP@.75
kraken-ft	45.22	95.37	45.48	57.16	47.16	**24.39**
doc-ufcn-ft	51.01	91.75	54.31	66.67	36.03	1.03
TESTR-ft	83.60	91.99	88.62	87.89	**51.66**	17.71

6 Analysis and Discussion

Fig. 2. Annotations sometimes miss the marginalia, which increases the false positive rate in models such as TESTR. **Left**: the annotation. **Right**: TESTR's prediction.

We analyse the experimental results and summarize them into the following key aspects.

1. Kraken works best on regular, dense, main-body text. It sometimes fail to capture the full line. It mostly fails to handle outliers. We also observe that Kraken achieve best number for the AP@.75 metric on both test sets. We argue it may be because Kraken use a comparably heavy post-processing algorithm to obtain the final bounding boxes.

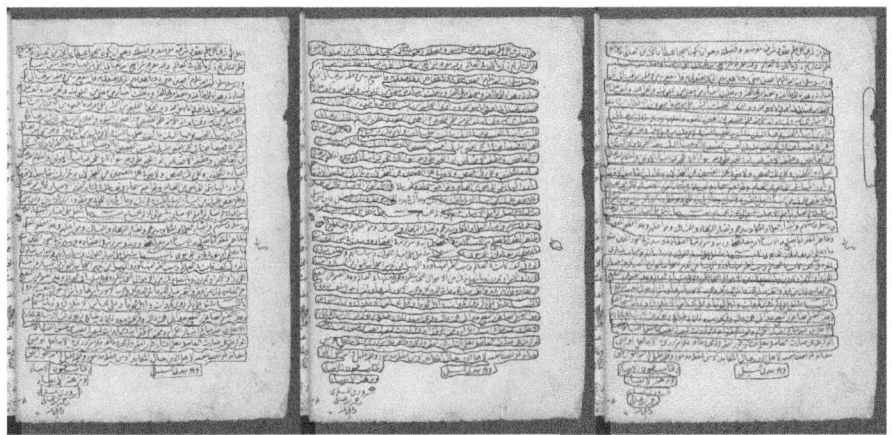

Fig. 3. Prediction of `aocp_ms_eval` (image id 47). From left to right: Kraken, DOC-UFCN, TESTR. DOC-UFCN is able to capture text lines in main text body, but the adjacent lines can not be differentiated.

2. DOC-UFCN performs very bad on iti-complex mainly because it cannot predict clean single lines when lines are irregular.
3. Overall, TESTR achieved the best performance in term of recall and AP@0.5. This indicates that TESTR is more robust to out-of-distribution new examples. As shown in Fig. 3 and Fig. 4, TESTR retrieved most of the marginalia text. However, we notice that the TESTR is expected to have uniform performance on similar text blocks on same page while TESTR usually drop one line. Another issue we found on TESTR is that the confidence of predicted polygons is low within $[0.1, 0.3]$ (shown in Fig. 5). Interestingly, the average confidence on `iti-complex` is significantly larger than that on `arabic_ms_eval`.
4. The CAR (Character Accuracy Rate) for DOC-UFCN is notably low. As observed, DOC-UFCN struggles to differentiate between closely spaced lines, often predicting adjacent lines as a single polygon. To address this issue, additional post-processing efforts could be implemented to enhance performance without necessitating modifications to the neural network model.
5. TESTR is approximately 8 times slower in training compared to the other two models, primarily due to its use of a Transformer-based backbone. Conversely, during inference, Kraken exhibits the slowest performance owing to its extensive post-processing requirements.

Fig. 4. Prediction of `aocp_ms_eval` (image id 46), From left t o right: Kraken, DOC-UFCN, TESTR. All three models predict most of the main-text body and Kraken works best on main-text body. TESTR recalls most of the text in marginalia.

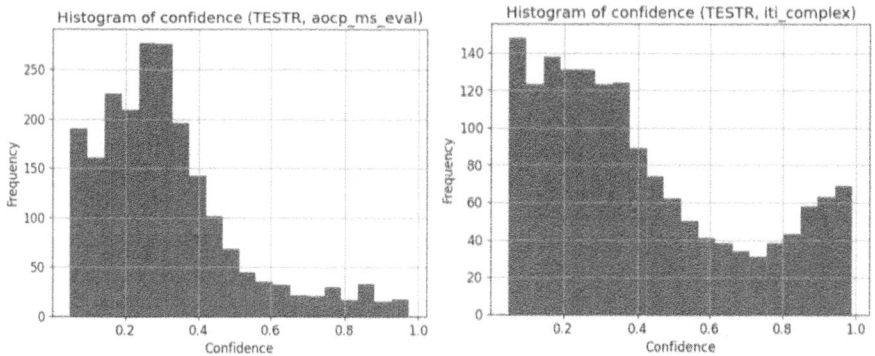

Fig. 5. Histograms of confidence scores of polygons predicted by TESTR on two test sets. The majority of the confidence scores are lower than 0.5.

7 Case Study: How Noisy Data Affects Performance

As we observed, the annotations in the training dataset, specifically in `arabic_ms_data`, are incomplete with approximately 7.5% of the main body text lines missing, and the missing rate for marginalia are doubled to 15.7%. Additionally, we note that the confidence scores for the predicted polygons from TESTR are lower than expected (shown in Fig. 5). These two observations prompt further investigation into how such noisy data influences model performance.

We also tested checkpoints before the model loss.

We randomly drop 20% and 50% of the polygons in `arabic_ms_data` before beginning the fine-tuning phase (phase II) and then evaluate the impact on test performance using both `aocp_ms_eval` and `iti-complex` datasets. As anticipated, there was a significant deterioration in performance with a 20% reduction in annotations. Surprisingly, however, performance did not decrease further when we removed an additional 30% of the annotations, as shown in Table 5. The `AP@.5` metrics barely changed when we reduced the number of annotated lines, when we manually look at some of the prediction.

Table 5. Ablation study on noisy data with a drop rate of 0 %, 20%, 50%.

drop rate	arabic_ms_data			iti-complex		
	0%	20%	50%	0%	20%	50%
AP@.50	**76.74**	73.11	70.98	**51.66**	43.32	43.19
AP@.75	**32.09**	18.24	18.00	**17.71**	13.76	13.58
avg. confidence	**31.36**	26.45	22.81	**40.67**	30.48	27.75

8 Conclusion

We empirically evaluate three major methods for text line extraction on Islamicate documents: baseline (`kraken`), region (`doc-ufcn`), and detection-based (`TESTR`). Scores for all three methods decrease drastically when tested exclusively on non-linear lines (i.e. the `iti-complex` dataset).

Evaluated at the object level, we find that a pixel-based model such as `kraken` works best on AP@0.75, while a detection-based method like `TESTR` performs better on AP@0.5. Indeed, `TESTR` achieves the highest performance for non-linear text line extraction in terms of recall and AP@0.5. Both `TESTR` and `kraken` significantly outperform `doc-ufcn` with respect to OCR CAR. Evaluated at the pixel level for F1, `TESTR` notably outperforms both `kraken` and `doc-ufcn`.

Our comparative evaluation reveals that each system has its unique strengths and weaknesses. Nevertheless, we believe these results also evidence the potential for transformer-based approaches to non-linear text line extraction.

References

1. Al-Barhamtoshy, H.M., Jambi, K.M., Abdou, S.M., Rashwan, M.A.: Arabic documents information retrieval for printed, handwritten, and calligraphy image. IEEE Access **9**, 51242–51257 (2021)
2. Al-Ma'adeed, S., Elliman, D., Higgins, C.: A data base for Arabic handwritten text recognition research. In: Proceedings Eighth International Workshop on Frontiers in Handwriting Recognition, pp. 485–489 (2002)
3. Baek, Y., Lee, B., Han, D., Yun, S., Lee, H.: Character region awareness for text detection. In: Proceedings of the IEEE/CVF Conference on Computer Vision and Pattern Recognition, pp. 9365–9374 (2019)

4. Barakat, B.K., Cohen, R., El-Sana, J.: VML-MOC: segmenting a multiply oriented and curved handwritten text line dataset. In: 2019 International Conference on Document Analysis and Recognition Workshops (ICDARW), vol. 6, pp. 13–18. IEEE (2019)
5. Boillet, M., Kermorvant, C., Paquet, T.: Multiple document datasets pre-training improves text line detection with deep neural networks. In: 2020 25th International Conference on Pattern Recognition (ICPR), pp. 2134–2141. IEEE Computer Society, Los Alamitos, CA, USA (2021). https://doi.org/10.1109/ICPR48806.2021.9412447, https://doi.ieeecomputersociety.org/10.1109/ICPR48806.2021.9412447
6. Boillet, M., Kermorvant, C., Paquet, T.: Multiple document datasets pre-training improves text line detection with deep neural networks. In: 2020 25th International Conference on Pattern Recognition (ICPR), pp. 2134–2141 (2021). https://doi.org/10.1109/ICPR48806.2021.9412447
7. Boillet, M., Kermorvant, C., Paquet, T.: Robust text line detection in historical documents: learning and evaluation methods. Int. J. Doc. Anal. Recogn. (IJDAR), pp. 1433–2825 (2022). https://doi.org/10.1007/s10032-022-00395-7
8. Douglas, D.H., Peucker, T.K.: Algorithms for the reduction of the number of points required to represent a digitized line or its caricature. Cartographica: Int. J. Geograph. Inf. Geovisual. **10**(2), 112–122 (1973)
9. Huang, M., et al.: Estextspotter: towards better scene text spotting with explicit synergy in transformer. In: Proceedings of the IEEE/CVF International Conference on Computer Vision (ICCV), pp. 19495–19505 (2023)
10. Kiessling, B.: The Kraken OCR system (2022). https://kraken.re
11. Lamsaf, A., Aitkerroum, M., Boulaknadel, S., Fakhri, Y.: Text line and word extraction of Arabic handwritten documents. In: Ben Ahmed, M., Boudhir, A.A., Younes, A. (eds.) Innovations in Smart Cities Applications, pp. 492–503 (2019)
12. Liao, M., Shi, B., Bai, X., Wang, X., Liu, W.: Textboxes: a fast text detector with a single deep neural network. In: Proceedings of the Thirty-First AAAI Conference on Artificial Intelligence, pp. 4161–4167 (2017)
13. Long, S., Ruan, J., Zhang, W., He, X., Wu, W., Yao, C.: TextSnake: a flexible representation for detecting text of arbitrary shapes. In: Ferrari, V., Hebert, M., Sminchisescu, C., Weiss, Y. (eds.) ECCV 2018. LNCS, vol. 11206, pp. 19–35. Springer, Cham (2018). https://doi.org/10.1007/978-3-030-01216-8_2
14. Lorigo, L., Govindaraju, V.: Segmentation and pre-recognition of Arabic handwriting. In: Eighth International Conference on Document Analysis and Recognition (ICDAR), pp. 605–609 (2005)
15. Lyu, P., Liao, M., Yao, C., Wu, W., Bai, X.: Mask TextSpotter: an end-to-end trainable neural network for spotting text with arbitrary shapes. In: Ferrari, V., Hebert, M., Sminchisescu, C., Weiss, Y. (eds.) Computer Vision – ECCV 2018. LNCS, vol. 11218, pp. 71–88. Springer, Cham (2018). https://doi.org/10.1007/978-3-030-01264-9_5
16. Mahmoud, S.A., et al.: Khatt: Arabic offline handwritten text database. In: 2012 International Conference on Frontiers in Handwriting Recognition, pp. 449–454 (2012)
17. Smith, D.A., Murel, J., Allen, J.P., Miller, M.T.: Automatic collation for diversifying corpora: commonly copied texts as distant supervision for handwritten text recognition. In: Computational Humanities Research Conference (CHR) (2023)
18. Tian, Z., Huang, W., He, T., He, P., Qiao, Yu.: Detecting text in natural image with connectionist text proposal network. In: Leibe, B., Matas, J., Sebe, N., Welling, M. (eds.) ECCV 2016. LNCS, vol. 9912, pp. 56–72. Springer, Cham (2016). https://doi.org/10.1007/978-3-319-46484-8_4

19. Wada, K.: LabelMe: image polygonal annotation with Python. https://doi.org/10.5281/zenodo.5711226, https://github.com/wkentaro/labelme
20. Ye, M., Zhang, J., Zhao, S., Liu, J., Du, B., Tao, D.: Dptext-detr: towards better scene text detection with dynamic points in transformer. In: Proceedings of the AAAI Conference on Artificial Intelligence, vol. 37, pp. 3241–3249 (2023)
21. Zahour, A., Likforman-Sulem, L., Boussalaa, W., Taconet, B.: Text line segmentation of historical Arabic documents. In: Ninth International Conference on Document Analysis and Recognition (ICDAR), vol. 1, pp. 138–142 (2007)
22. Zahour, A., Taconet, B., Mercy, P., Ramdane, S.: Arabic hand-written text-line extraction. In: Proceedings of Sixth International Conference on Document Analysis and Recognition, pp. 281–285 (2001)
23. Zhang, X., Su, Y., Tripathi, S., Tu, Z.: Text spotting transformers. In: Proceedings of the IEEE/CVF Conference on Computer Vision and Pattern Recognition (CVPR), pp. 9519–9528 (2022)

A New Framework for Error Analysis in Computational Paleographic Dating of Greek Papyri

Giuseppe De Gregorio[1(✉)], Lavinia Ferretti[1], Rodrigo C. G. Pena[1], Isabelle Marthot-Santaniello[1], Maria Konstantinidou[2], and John Pavlopoulos[3]

[1] University of Basel, Basel, Switzerland
{giuseppe.degregorio,lavinia.ferretti,rodrigo.cerqueiragonzalezpena,
i.marthot-santaniello}@unibas.ch
[2] Democritus University of Thrace, Komotini, Greece
mkonst@helit.duth.gr
[3] Athens University of Economics and Business, Athens, Greece
annis.pavlo@aueb.gr

Abstract. The study of Greek papyri from ancient Egypt is fundamental for understanding Graeco-Roman Antiquity, offering insights into various aspects of ancient culture and textual production. Palaeography, traditionally used for dating these manuscripts, relies on identifying chronologically relevant features in handwriting styles yet lacks a unified methodology, resulting in subjective interpretations and inconsistencies among experts. Recent advances in digital palaeography, which leverage artificial intelligence (AI) algorithms, have introduced new avenues for dating ancient documents. This paper presents a comparative analysis between an AI-based computational dating model and human expert palaeographers, using a novel dataset named Hell-Date comprising securely fine-grained dated Greek papyri from the Hellenistic period. The methodology involves training a convolutional neural network on visual inputs from Hell-Date to predict precise dates of papyri. In addition, experts provide palaeographic dating for comparison. To compare, we developed a new framework for error analysis that reflects the inherent imprecision of the palaeographic dating method. The results indicate that the computational model achieves performance comparable to that of human experts. These elements will help assess on a more solid basis future developments of computational algorithms to date Greek papyri.

Keywords: Greek papyri · Computational dating · Palaeography · Error analysis · Human comparison

1 Introduction

Greek papyri preserved thanks to the dry climate of Egypt represent an unparalleled primary source for the study of Graeco-Roman Antiquity. These texts, which cover a wide range of contents from documentary testimonies (contracts, letters...) to literary works, play a crucial role in our understanding of ancient culture in general and of handwriting evolution and book production in particular. A rough classification of papyri divides them between documentary and literary texts: while the former are usually written in fast, informal cursive scripts, the latter use formal scripts, i.e. slow, detached and easy-to-read, sometimes calligraphic scripts which received the designation of "book hands". Regardless of the type of papyrus, the informational value of such manuscripts significantly increases when their dating can be established. Sometimes, it is possible to assign them a date thanks to textual or archaeological evidence; when such clues are lacking, researchers resort to palaeography to estimate an approximate and broad date for their production.

Palaeographers operate under the assumption that writing styles share some features among coeval specimens and gradually change over time. For dating, scholars use a comparative technique: a date is proposed by comparison with other previously dated samples, preferably but not necessarily with objectively dateable papyri. Although multiple palaeographers identify some specific writing features as chronologically significant, to date, no unified methodology is consensual among experts. This results in a situation where each palaeographer is free to focus on aspects of writing they deem most significant. Moreover, even when common features can be identified, they are rarely objectively measurable or calculable [5,6,12,15,16]. This palaeographical method, heavily relying on personal expertise that can be acquired from a long acquaintance with manuscripts, is open to many uncertainties, if not errors, and is indisputably difficult to communicate. This explains the regular occurrence of conflicting results among different experts [7,12,16].

Recently, digital palaeography has introduced sophisticated techniques, including image analysis using artificial intelligence (AI) algorithms. However, interpreting the results obtained from such techniques requires caution, as the features used for making a decision in trained models may differ from those relevant to human experts.

In this work, we aim to provide a dating framework for Greek papyri that incorporates the inherent imprecision of chronological attribution only based on handwriting. To this end, we have:

1. compiled a new dataset, named Hell-Date, composed of images of papyri whose exact year of writing is established thanks to unequivocal internal evidence;
2. evaluated the performance of a convolutional neural network trained on visual inputs, specialised in dating ancient documents when applied to a dataset characterised by precise and granular dating (Hell-Date);
3. set an experiment based on the network pipeline to evaluate the results of the model compared with that of expert scholars;
4. adopted metrics that integrate the chronological imprecision inherent to the method to analyse the error.

2 Related Works

Various techniques have been applied in the past to the challenge of dating ancient documents using computational methods, changing according to factors such as document type, language, and historical period.

In the literature, diverse standard Machine Learning techniques have been tried. Dhali et al. [10] used a Support Vector Regressor-based technique to date a collection of 595 Dead Sea Scrolls written in the Hebrew alphabet, ranging from 250 to 135 BCE, through feature extraction from manuscript scripts. Adam et al. [1] proposed a sparse representation-based method to date historical handwritten Arabic manuscripts, employing a K-nearest neighbour (KNN) approach. Some methodologies rely on textual analysis for dating, as Baledent et al. [3], who introduced a dataset that features numerous ancient documents in French and used decision trees and random forests at both character and token levels.

Noteworthy advancements have emerged from techniques rooted in convolutional networks and Deep Learning models. Li et al. [13] presented an approach using convolutional neural networks (CNNs) alongside text features extracted via optical character recognition for estimating the publication date of historical English printed documents from the 15th to the 19th century. Cloppet et al. [8] and Seuret et al. [21] propose competitions whose tasks include dating manuscripts and printed material from the medieval ages. Hamid et al. [11] leveraged pre-trained CNNs to date images of medieval Dutch cards from the 14th to the 16th century, with a focus on image fragments. Wahlberg et al. [23] proposed a deep learning methodology for dating pre-modern handwritten documents, achieving results comparable to human experts on a dataset comprising over 10,000 medieval Swedish cards.

While various methodologies have been explored across different languages and historical periods, fewer studies have focused on texts written on papyrus in ancient Greek. Pavlopoulos et al. [19] experimented on dating Greek papyri analysing textual contents using regression methodologies. One of the first approaches to dating Greek papyrus images via convolutional networks was taken by Paparrigopoulou et al. [17], reporting an average dating error of more than a century. Subsequent work by Pavlopoulos et al. [20], focusing on literary papyri from the Roman period (1st - 4th CE), demonstrated improved accuracy through a segmentation strategy. The authors showed that segmenting to line level rather than working at the whole document level reduces the average dating error.

3 The Methodology

3.1 Datasets

The New Hell-Date Dataset. Hell-Date is a dataset composed of 187 images of 155 papyri written in Greek and dated to the Hellenistic period (from the late 4th to the 1st c. BCE)[1]. These texts are securely dated, i.e. they contain textual

---

[1] The dataset is accessible at the following link: https://d-scribes.philhist.unibas.ch/en/hell-date/.

evidence that points to the exact year in which they were written - usually the mention of the date somewhere in the text. For instance, the papyrus TM 244 shown in Fig. 1 is a contract in which the first 3 lines of column 2 explicitly mention that it was written in the 17th year of reign of King Ptolemy Alexander and Queen Berenice, the fourth day of the month Mecheir[2]. This date, converted into the modern system, corresponds to February 17, 97 BCE.

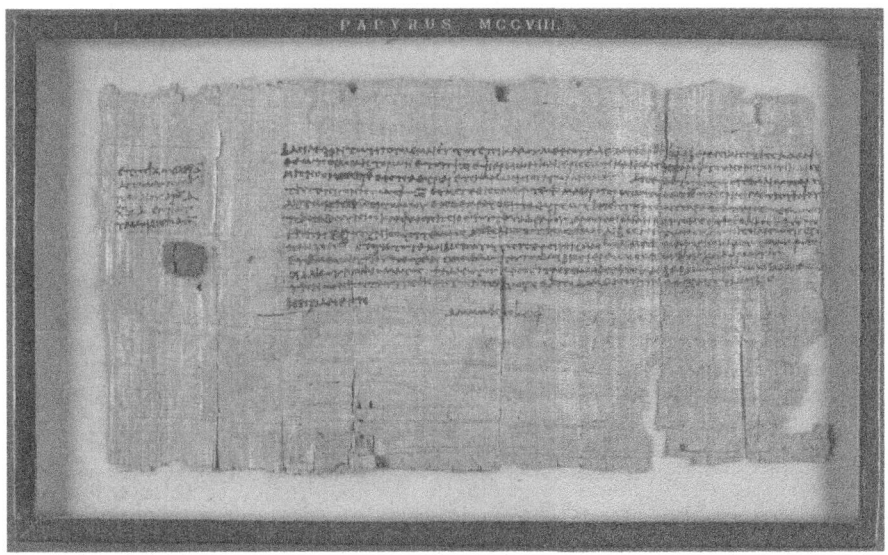

Fig. 1. Image of the papyrus TM 244 = P.Lond. III 1208 = P.Lond. inv. 1208. Image courtesy of the British Library. The first 3 lines of column 2 translate: "*When Ptolemy, also known as Alexander, and Berenice, his sister, the mother-loving gods, are reigning for the 17th year, under the priests and priestesses and the canephoros which are in charge, day 4 of the month Mecheir, in Pathyris, under the notary Ammonios*".

Several dating systems are used in Hellenistic papyri, and none, as expected, strictly aligns with our modern Gregorian calendar [2,4,18]. It is thus impossible to give one Gregorian year of writing for some papyri; for instance, papyrus TM 121853 was written in the month of Audnaios of the 18th year of reign of king Ptolemy II, which corresponds to a period from late December 268 to late January 267 BCE. In these cases, less than twenty in the full dataset, we have given both years in the dataset metadata but used arbitrarily the oldest year for the purposes of this paper.

Restricted to the Hellenistic period (from -310 to -3), Hell-Date was selected on purpose to have an even distribution with about 50 documents per century. These texts are written in so-called "documentary hands", i.e. fast scripts that

---

[2] Papyri are cited according to Trismegistos (TM) Numbers, for which cf. [9] and https://www.trismegistos.org/about_how_to_cite.php.

prioritise speed over legibility and aesthetics. Such scripts are often cursive, present ligatures and can be hard to read.

The images incorporated within the dataset originate from diverse online collections and resources undergoing digitisation over a span exceeding two decades, using distinct imaging protocols that are not documented[3]. Consequently, they exhibit considerable property variations, notably about scaling to actual size, colour capture, resolution, and bit depth. Moreover, due to the condition of the original papyrus, these images may exhibit noise, including surface damage, empty spaces, and delineations marking the papyrus edges. An example of such problems can be seen in Fig. 1, where there are stains and holes on the surface, empty spaces between columns and a modern frame with modern writing. All the collected images were segmented into lines of text using the docExtractor segmentation tool [14] followed by a manual correction. Subsequently, lines longer than twice the average length were further divided, while short lines that contained only a few characters or single words were discarded. This process resulted in clippings of text lines with comparable dimensions, without any layout information such as column width or document size, and yielded a total of 8,230 images of text lines, with a peak of distribution in the late 2nd c. BCE.

Comparable Dataset. For comparison, we selected from the existing literature the Papyri Literary Lines (PLL) dataset [20]. To the best of our knowledge, this is the only existing dataset tailored for the task of dating Greek papyri, in this case, from the Roman period. It is composed of 2,774 line images extracted from 159 images of Greek papyri, selected from the *Collaborative Database of Dateable Greek Bookhands*; for details on the PLL dataset, one could refer to the original publication.

The PLL dataset is different from Hell-Date. These papyri date between the 1st and the 4th c. CE. Their date can only be estimated with some approximation; therefore, it is only given at a century level. The chronological distribution is uneven, with a peak in the 2nd and 3rd c. CE. They are written in book hands, i.e. formal scripts usually used for literary texts. These hands may present artificial phenomena like archaism or fossilisation that can complicate their dating. Last, the PLL dataset contains almost three times fewer line images than Hell-Date. The characteristics of the two datasets are summarised in Table 1.

For the experiments, the datasets were partitioned such that 90% of the document images were allocated to the training set, while the remaining 10% were designated for the test set. Importantly, this division was performed at the level of document image rather than the text line. This approach ensures that no text line present in the test set originates from a document image used for training. Following the construction of the test set, a validation set comprising 10% of the training data is extracted. At the end, the Hell-Date training set comprises 7424 line images from 169 documents and the test set of 806 lines

---

[3] For a list of these resources, see the online description of the dataset: https://d-scribes.philhist.unibas.ch/en/hell-date/dataset/.

from 18 documents. Conversely, the PLL training set is composed of 2496 line images from 146 documents, and the test set of 270 lines from 13 documents.

Table 1. Datasets overview.

	Hell-Date	PLL
Period	Hellenistic	Roman
Centuries	4th-1st BCE	1st-4th CE
Precision of Date	to the Year	to the Century
Type of Writing	Cursive Hand	Book Hand
N Documents	187	159
N Lines	8230	2774

3.2 The Computational Dating

The Model. The convolutional network used in this study is referred to as fCNN, originally introduced by Pavlopoulos et al. [20][4]. The fCNN network is presented in two versions: fCNNc as a classifier and fCNNr as a regressor. Employing the model as a classifier requires a rigid discretisation of the time axis to define distinct classes. However, the discretisation of time implies the loss of granular information on exact dates. This conflicts with the primary aim of achieving the most precise dating feasible, particularly considering the granularity of chronological ground truth in the Hell-Date dataset. Therefore, the fCNNr regressor version aligns well with our goals.

The model architecture consists of two Conv2D layers featuring 32 and 64 channels respectively, followed by a 3-layer feed-forward neural network culminating in a single output neuron responsible for date estimation. Convolutional operations are defined by a kernel size of 5, stride of 1, zero padding, and subsequent maximum pooling with a 2×2 window. The feed-forward component processes a flattened representation obtained from the convolutional layers, sequentially reducing the neuron count to 1024 and then 512 before the final date prediction. Each layer incorporates Rectified Linear Unit (ReLU) activation functions. Functioning as a regression model, this network takes as input the image of a handwritten text line and produces a single numerical output representing the inferred date.

During the model training phase, data augmentation techniques are employed to enhance network robustness. This involved random deletion of image fragments with a probability of 0.5, replacing the pixel values with 0.5. Additionally, images undergo transformations including Gaussian blur with a kernel size of 3

---

[4] The code published in the article is available at https://github.com/ipavlopoulos/palit.

and random affine transformations up to 3rd degree. Furthermore, random cropping and resizing are applied to each image while maintaining a 1:6 aspect ratio to minimise excessive alteration of the manuscript content within the images.

Bearing in mind that the Hell-Date dataset provides a more granular ground truth than PLL, we experimented with several scenarios with various combinations of Hell-Date and PLL to evaluate the relevance of transfer learning on the performance of computational dating.

First, we experimented with training and testing the network on Hell-Date alone and on PLL alone. Second, initial training uses one dataset, and further training uses the other one, testing on the last one. This means, in the first case we train first on PLL and then we continue on Hell-Date and test on Hell-Date, and in the second case, we train on Hell-Date, then on PLL and test on PLL. Finally, we first merged the Hell-Date and the PLL training sets and then, we trained the network on this combination. We tested on both test sets and on the combination of test sets. As the PLL and Hell-Date datasets differ in size, we explored the partial use of the larger dataset to maintain data balance in the union dataset.

3.3 The Human Experts Dating

Expert scholars in the study of ancient manuscripts can ascertain the approximate age of documents based on the palaeographic analysis of handwritten text fragments. In this study, we recruited five highly experienced scholars specialising in Hellenistic Greek papyri to participate in a dating assessment.

To arrange the experiment, we selected three lines from each document within the Hell-Date test set. This number allows, on the one hand, reflecting the variety and average preservation of the original document and, on the other hand, not requiring too much time from the participants. We carefully selected lines that do not contain any textual clue on the date of the text. We created a form presenting the images in random order, asking the respondents to provide for each line a dating interval expressed as two integers (starting and ending years of the interval). The respondents were asked to be as specific in their dating as they felt confident and to base themselves only on the appearance of the writing. They only had one chance to fill out the form and were given thirty minutes to answer, so that they could not try to identify in the literature the text from which the line came. They were not asked to justify their dating. Although dating based on lines is uncommon in traditional palaeography, the experiment was devised as such to allow for comparability with the results of the computational model.

4 Results

In this section, we present the results of our experiments, beginning with the results obtained through the computational dating method. Subsequently, we provide the results obtained by the human experts.

4.1 Computational Dating

For training the various models, we employed Adam optimisation with a learning rate of 1e−3 and a batch size of 16. Training iterations extended up to 200 epochs, implementing an early stopping policy with patience of 20 epochs. The loss function employed throughout the training was the mean squared error. Regarding representation, floating point numbers are chosen to represent centuries, such that the integer part represents the exact year (e.g. 1.50 corresponds to the year 150). Consequently, Mean Absolute Error (MAE) and Mean Squared Error (MSE) are adopted as the evaluation metric for error assessment.

Table 2 presents the Mean Absolute Error (MAE) and the Mean Squared Error (MSE) when the network is trained and tested on the same dataset. Given the larger dimension of the Hell-Date dataset compared to PLL, training experiments were repeated with Hell-Date while considering different percentages of the training set each time. As can be observed from the error reported in the table, the network achieves a MAE of approximately 55 years for both datasets. However, these outcomes are attained solely when using 100% of the available data for training. Indeed, when the size of the training data between PLL and Hell-Date is comparable (i.e. when approximately 35% of the Hell-Date training set is considered), a larger error is noted on Hell-Date.

Table 2. Number of text line images for training, validation, test and results in terms of MAE and MSE.

	PLL 100%	Hell-Date 100%	Hell-Date 50%	Hell-Date 35%	Hell-Date 10%
Trainings Lines	2496	7424	3340	2338	667
Validation Lines	270	806	372	260	75
Test Lines	270	806	806	806	806
MAE	**0.5418**	**0.5637**	0.5923	0.6050	0.6446
MSE	**0.5161**	**0.4922**	0.5274	0.5776	0.5965

Can Transfer Learning Improve Performances? At this point, we proceed to assess the feasibility of transferring learning between the two datasets. In this context, our approach involves initially training the network on one of the two distinct datasets before proceeding with additional training on the other dataset. Table 3 displays the results achieved on the Hell-Date test set after training conducted initially on PLL followed by training on Hell-Date. Table 4 exhibits the results on the PLL test set when the network undergoes training initially with Hell-Date followed by training with PLL. Comparing the last tables with the results in Table 2, it becomes evident that the transfer of learning yielded marginal improvements in dating the Hell-Date test set, leading to a slight decrease in both the MAE and the MSE. Conversely, there was a decline in performance when evaluating the PLL documents.

Table 3. Results in terms of MAE and MSE on Hell-Date test set when the network is trained on PLL and further trained on a percentage of Hell-Date.

	Hell-Date 0%	Hell-Date 25%	Hell-Date 50%	Hell-Date 75%	Hell-Date 100%
MAE	9.9426	0.6452	0.6390	**0.5357**	0.5424
MSE	133.88	0.6568	0.6669	0.6669	**0.4883**

Table 4. Results in terms of MAE and MSE on PLL test set when the network is trained on Hell-Date and further trained on a percentage of PLL.

	PLL 0%	PLL 25%	PLL 50%	PLL 75%	PLL 100%
MAE	96.089	0.6505	0.5904	0.5857	**0.5687**
MSE	51342	0.6315	0.5689	0.5664	**0.5362**

Finally, we attempt to train the network by combining the two datasets. In this experiment, training is conducted on a set comprising the union of the training sets from the different datasets. Subsequently, the trained network is assessed on the individual test sets of the two datasets as well as on the combined test set. Given the considerable size discrepancy between the Hell-Date and PLL test sets, we iterate the training process while adjusting the percentage of Hell-Date used for fusion; the two training sets are fairly balanced when considering approximately 35% of the Hell-Date training set. The outcomes of this experiment are summarised in Table 5.

Table 5. Results in terms of MAE and MSE when the network is trained on the union between PLL and Hell-Date

PLL %	Hell-Date %	Test Set PLL		Test Set Hell-Date		Test Set PLL + Hell-Date	
		MAE	MSE	MAE	MSE	MAE	MSE
100	100	1.0612	2.2559	**0.7918**	**1.4973**	**0.8756**	**1.7332**
100	50	0.8249	1.3885	0.9097	1.9220	0.8834	1.7560
100	35	0.9186	1.6376	0.9010	1.7588	0.9095	1.7211
100	10	**0.7568**	**0.8545**	1.2530	3.3185	1.0987	2.5512

As depicted in Table 5, the percentage used for Hell-Date significantly impacts the results. The optimal results on the combined test sets are achieved when employing 100% of both training sets, on the other hand, it is evident that the error distribution is uneven across the two datasets. Notably, when the two training sets are balanced in terms of size (approximately 35% of Hell-Date), the errors on the two test sets are close. By comparing the results of Tables 2, 3, 4, and 5, it appears that merging the datasets does not bring any improvement: the results are even the worst we obtained in all our experiments. The different nature of the handwriting in the two datasets may explain this behaviour.

4.2 The Human Experts Dating

As explained above, five expert scholars were provided with a form to date 54 line images (3 lines for each of the 18 images of the Hell-Date test set)[5]. Some experts opted not to provide answers for specific images presented in the form. To compute the MAE, for these unanswered entries, the error value corresponding to the maximum error within the dataset was used as an answer. Table 6 presents the MAE observed in the responses of each expert, along with the average error calculated from the collective performances. Additionally, the table shows the error calculated by disregarding the rows where the expert did not submit a response.

Table 6. MAE obtained from the dating of the expert scholars, by including and excluding empty answers.

Expert	01	02	03	04	05
MAE	1.27	0.41	2.28	0.48	0.62
MAE (excl. empty answers)	1.13	0.41	1.75	0.48	0.53
# empty answers	3	0	20	0	2

Given the variability in errors among different experts, one may wonder about the extent of agreement across these experts. To address this inquiry, Table 7 presents some indices to measure the level of agreement between experts: the Mean Pairwise MAE, the Mean Pairwise Spearman and Pearson Correlation, and the Fleiss' kappa index. The first is computed by calculating the MAE considering the dates of one expert as truth and those of the others as predictions. The second and third compute the correlations between any two experts and average them. Finally, to compute the kappa index, it was necessary to discretise the responses to enable the assignment of shared labels. To achieve this, the time axis was discretised with a step size of 25 years. Subsequently, for each document, a positive (1) or negative (0) label was assigned to each interval of the discretised time axis based on the expert's prediction, such that intervals covered partially or totally by the prediction were set to 1, and intervals fully outside the prediction to 0. As can be seen from the table, we can record a very slight agreement among the experts, being the Mean Pairwise MAE high, the Mean Pairwise correlations indices close to 0.5, and Fleiss' Kappa slightly greater than zero.

---

[5] The values for individual respondents are accessible at the following link: https://d-scribes.philhist.unibas.ch/en/hell-date/.

5 Discussion

5.1 Error Analysis

In the context of automatic dating of historical documents, error is a significant challenge to address. Analysis based on Mean Absolute Error may be limited, as a writing style does not correspond to a single precise date, but rather to a time interval that reflects the gradual process of handwriting evolution. To address this challenge, we employed the method of the Error Time Window (ETW). The ETW is defined as a rectangular window function Π, centred around the production year of the document Y, characterised by a certain width α in terms of time t expressed in years:

$$ETW = \Pi(\alpha t - Y) \tag{1}$$

Using this window, we evaluate whether a prediction obtained by the neural network falls inside or outside this range.

This approach enables the calculation of the Accuracy index ($A(\alpha)$), representing the proportion of correct predictions compared to the total predictions made for a given α width of the ETW:

$$A(\alpha) = \frac{P}{P+N} \tag{2}$$

Here, P (positive) represents correct predictions for each line, and N (negative) indicates incorrect predictions relative to the defined time window. Figure 2 illustrates the accuracy trend as the size α of the ETW varies, considering the model that performed best on the dating of the Hell-Date test set, which is the one trained first on PLL then on Hell-Date.

We can interpret the results in terms of MAE presented in Table 3 by focusing on the case where the ETW width equals the MAE. With an approximate MAE of 55 years, the relative ETW spans approximately 110 years. Under this time tolerance, the network correctly dates nearly 60% of the test documents. Notably, Fig. 2 shows that when the ETW width decreases to 50 years ($\alpha = 0.5$), the accuracy remains relatively stable at around 50%. This suggests that despite the average error, the predictions are accurate for a significant proportion of documents, so we can expect the network to give accurate estimations of the century of production of handwriting.

Table 7. Indices to measure the agreement among experts.

Index	Value
Mean Pairwise MAE	105.91
Mean Pairwise Spearman Corr	0.54
Mean Pairwise Pearson Corr	0.53
Fleiss' Kappa	0.03

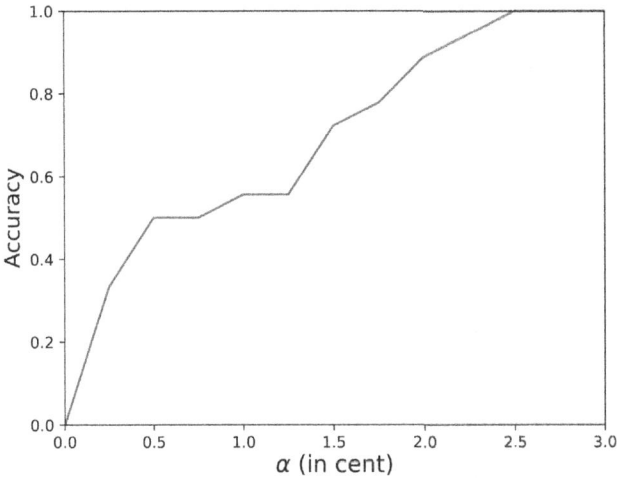

Fig. 2. Accuracy of the Hell-Date test set dating according to the size α of the ETW.

Looking further at Fig. 2, we observe that the network achieves an accuracy of 1 only with an error window width equal to two and a half centuries ($\alpha = 2.5$). Given the vast breadth of this time interval, it is worth analysing further the error distribution of the network predictions for single lines. The left plot in Fig. 3 shows the time-axis distribution of the predictions for individual lines of each document in the test set. Documents are chronologically ordered, starting on the top with the most ancient. While some documents exhibit less precise dating, over half of the test cases display very accurate average predictions. Additionally, there appears to be a trend towards greater precision when dating in the second half of the 2nd c. BCE.

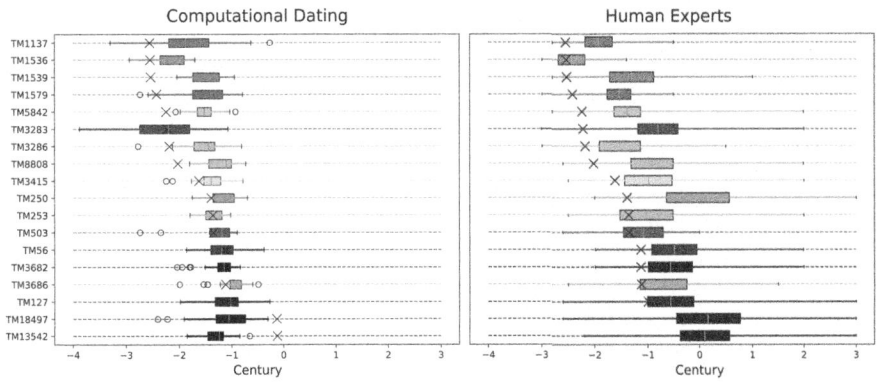

Fig. 3. Datings by fCNN (on the left) and the experts (on the right) a box plots per document. The ground truth (actual date of writing) is indicated with an X.

5.2 Human-AI Comparison

To compare the outcomes achieved by both the expert scholars and the computational model, we use the MAE values obtained by the best-performing model and the ones of humans without excluding empty answers. To facilitate comparison, we report in the right plot of Fig. 3 the error distribution made by expert scholars for each document. Table 8 sums up the comparison, giving the MAE and standard deviation σ of the predictions for each document, as committed by both the computational model and the human experts. With the exception of TM 1137 and 1536, the two most ancient documents in the test set, the computational approach yields better results in terms of MAE while keeping a lower standard deviation.

However, the results of the individual experts varied heavily. Therefore, we produced Fig. 4 summarising all results in terms of MAE. For the sake of thoroughness, we included both MAE values with and without empty answers. It is noticeable that while the computational model demonstrates superior performance compared to the average human result, only two experts achieved sur-

Table 8. Results (MAE and standard deviation σ) of the prediction for each document for the AI model, the mean human experts, and the best-performing human expert.

TM	AI-Model		Human Experts		Human Expert H2	
	MAE	σ	MAE	σ	MAE	σ
TM1137	0.78	0.57	0.70	0.53	0.15	0.19
TM1536	0.47	0.34	0.14	0.34	0.21	0.47
TM1539	1.06	0.32	1.29	0.98	0.58	0.42
TM1579	0.94	0.39	1.09	0.61	0.24	0.25
TM5842	0.72	0.28	1.07	1.42	0.10	0.17
TM3283	0.52	0.65	1.59	1.43	0.09	0.24
TM3286	0.71	0.39	0.81	0.82	0.20	0.25
TM8808	0.84	0.28	1.27	1.28	0.29	0.50
TM3415	0.32	0.31	0.74	0.91	0.09	0.36
TM250	0.30	0.29	1.40	1.29	0.58	0.87
TM253	0.17	0.20	0.75	1.09	0.28	0.40
TM503	0.42	0.62	0.73	0.67	0.27	0.53
TM3682	0.19	0.29	1.03	1.10	0.39	0.43
TM56	0.25	0.31	0.91	1.20	0.37	0.60
TM3686	0.23	0.21	0.77	0.96	0.35	0.54
TM127	0.29	0.37	1.05	1.40	0.61	0.67
TM18497	0.92	0.43	1.39	1.74	1.17	0.75
TM13542	1.16	0.27	1.43	1.33	1.41	0.41

passing the AI. The confidence of these experts is further visible by the fact that they did not leave any empty answer.

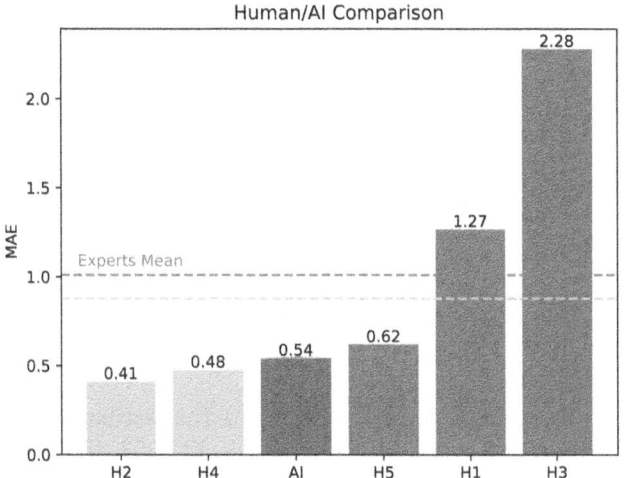

Fig. 4. Comparison between human experts and AI Model performance.

At this point, it is relevant to directly compare the results of the computational model with those from the best expert (H2). Figure 5 shows the error distribution for each. In general, the expert tends to give larger estimations than the AI, as can also be seen for the σ and MAE values reported in Table 8. Therefore, the ground truth falls in the two best quartiles in twelve cases compared to six cases for the computational model. This fact suggests that even the best-performing human tends to give a larger time span in order to increase the probability that the ground truth falls in the given range. Moreover, the human has a less marked tendency to date papyri in the 2nd c. BCE than the computational approach.

To look more deeply at the differences, we had a look at some of the papyri in the test set. The best-performing human dates better documents from the 3rd c. BCE. Compared to them, fCNNr poorly performed in dating these early documents and tended to attribute them to the following century. Although not all humans outperformed the AI model in dating documents from the 3rd c. BCE, it is clear from the human mean that experts usually correctly date one document: TM 1536. This papyrus is written in a style that is well studied in papyrological scientific literature, the so-called Alexandrian chancery style [5, 26–31], a fact that could explain the good results of human experts. The machine outperforms the best human for documents dated in the late 2nd c. BCE, TM 503, 253 and 250. These three papyri come from an ensemble of papyri penned by the same two writers, a man called Dryton and his son Esthladas (see [22]). Papyri from the same ensemble are present in the training set. The high precision of the

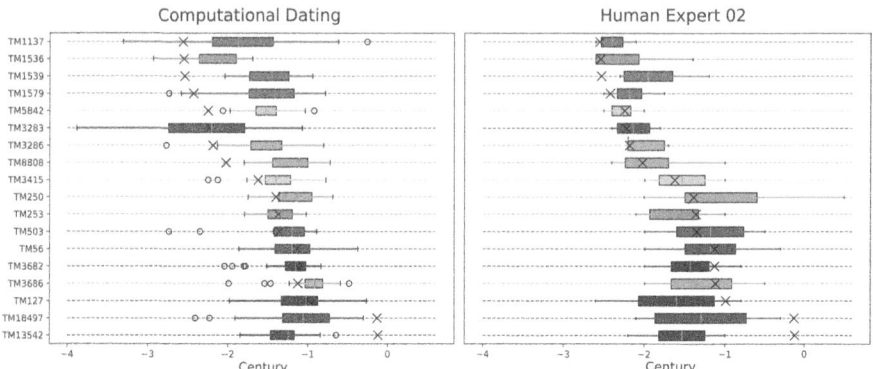

Fig. 5. Comparison between computational model performance (left) and best-performing Human Expert (right).

model in dating these three documents suggests that scribal identification may play a role in the dating process. In the case of TM 253, the papyrus is today divided into two fragments, one preserved in London, the other in Heidelberg. The image of the London fragment was in the training set, and the image of the Heidelberg one was in the test set. One may suggest that, despite the difference in preservation and digitisation of the two fragments (for instance, in the colour scale and brightness of the two images), the network managed to identify identical scripts which led to a relatively precise prediction. Finally, for the two most recent documents, TM 13542 and 18497, both the fCNNr network and the best-performing human wrongly located them in the middle of the possible time span. On the contrary, the average answer of human experts correctly dated these documents. However, we noticed high uncertainty of the human predictions and strong disagreement among the experts concerning these two documents. Among our respondents, some had only expertise in Hellenistic papyri and some in Hellenistic and Roman papyri. As these very late documents' writing is close to Roman cursive scripts, the experts with knowledge of Roman scripts may have been facilitated in dating those pieces compared to those with exclusively Hellenistic expertise.

6 Conclusion

In this study, we introduced a novel, precisely dated dataset named Hell-Date. We used fCNN to predict dates and observed only little improvement using transfer learning. To evaluate the performance of dating models in the context of historical documents, we propose a framework that incorporates the inherent imprecision of the palaeographic dating method. It relates the accuracy of the prediction with the size variation of the Error Time Window and integrates for each document the variability of the prediction, providing an in-depth view of the predictive capabilities of the model. The comparison with human results shows that current models are already able to give results comparable to those of experts. Future works will aim on the one hand at covering a larger time period with a more numerous dataset and on the other hand at better defining handwriting style similarities to provide new interpretations of their variety.

References

1. Adam, K., Baig, A., Al-Maadeed, S., Bouridane, A., El-Menshawy, S.: Kertas: dataset for automatic dating of ancient Arabic manuscripts. Int. J. Doc. Anal. Recogn. (IJDAR) **21**, 283–290 (2018)
2. Bagnall, R.S.: Practical help: chronology, geography, measures, currency, names, prosopography, and technical vocabulary. In: Bagnall, R.S. (ed.) The Oxford Handbook of Papyrology, pp. 179–196. Oxford University Press, Oxford (2009)
3. Baledent, A., Hiebel, N., Lejeune, G.: Dating ancient texts: an approach for noisy French documents. In: Sprugnoli, R., Passarotti, M. (eds.) Proceedings of LT4HALA 2020 - 1st Workshop on Language Technologies for Historical and Ancient Languages, pp. 17–21. European Language Resources Association (ELRA), Marseille, France (2020). https://aclanthology.org/2020.lt4hala-1.3
4. Bennett, C.: Alexandria and the Moon: An Investigation into the Lunar Macedonian Calendar of Ptolemaic Egypt. Peeters, Leuven (2011)
5. Cavallo, G.: La scrittura greca e latina dei papiri: una introduzione. F. Serra, Pisa, Rome (2008)
6. Cavallo, G.: Greek and Latin writing in the papyri. In: Bagnall, R.S. (ed.) The Oxford Handbook of Papyrology, pp. 101–148. Oxford University Press, Oxford (2009)
7. Choat, M.: Dating papyri: familiarity, instinct and guesswork. J. Study New Testament **42**(1), 58–83 (2019). https://doi.org/10.1177/0142064X19855580
8. Cloppet, F., Eglin, V., Helias-Baron, M., Kieu, C., Vincent, N., Stutzmann, D.: ICDAR2017 competition on the classification of medieval handwritings in latin script. In: 2017 14th IAPR International Conference on Document Analysis and Recognition (ICDAR), vol. 01, pp. 1371–1376 (2017). https://doi.org/10.1109/ICDAR.2017.224
9. Depauw, M., Gheldof, T.: Trismegistos: an interdisciplinary platform for ancient world texts and related information. In: Bolikowski, Ł, Casarosa, V., Goodale, P., Houssos, N., Manghi, P., Schirrwagen, J. (eds.) TPDL 2013. CCIS, vol. 416, pp. 40–52. Springer, Cham (2014). https://doi.org/10.1007/978-3-319-08425-1_5
10. Dhali, M.A., Jansen, C.N., de Wit, J.W., Schomaker, L.: Feature-extraction methods for historical manuscript dating based on writing style development. Pattern Recogn. Lett. **131**, 413–420 (2020)
11. Hamid, A., Bibi, M., Moetesum, M., Siddiqi, I.: Deep learning based approach for historical manuscript dating. In: 2019 International Conference on Document Analysis and Recognition (ICDAR), pp. 967–972. IEEE (2019)
12. Harrauer, H.: Handbuch der griechischen Paläographie. A. Hiersemann, Stuttgart (2010)
13. Li, Y., Genzel, D., Fujii, Y., Popat, A.C.: Publication date estimation for printed historical documents using convolutional neural networks. In: Proceedings of the 3rd International Workshop on Historical Document Imaging and Processing, pp. 99–106 (2015)

14. Monnier, T., Aubry, M.: docextractor: an off-the-shelf historical document element extraction. In: 2020 17th International Conference on Frontiers in Handwriting Recognition (ICFHR), pp. 91–96 (2020). https://doi.org/10.1109/ICFHR2020.2020.00027
15. Nongbri, B.: Palaeographic analysis of codices from the early Christian period: a point of method. J. Study New Testament **42**(1), 84–97 (2019). https://doi.org/10.1177/0142064X19855582
16. Orsini, P., Clarysse, W.: Early new testament manuscripts and their dates: a critique of theological palaeography. Ephemer. Theol. Lovan. **88**, 443–474 (2012). https://doi.org/10.2143/ETL.88.4.2957937
17. Paparrigopoulou, A., Kougia, V., Konstantinidou, M., Pavlopoulos, J.: Greek literary papyri dating benchmark. In: Coustaty, M., Fornés, A. (eds.) ICDAR 2023. LNCS, vol. 14193, pp. 296–306. Springer, Cham (2023). https://doi.org/10.1007/978-3-031-41498-5_21
18. Paulissen, J., Vandorpe, K.: Dating early Ptolemaic salt tax receipts: the Egyptian tax year. Zeitschrift für Papyrologie und Epigraphik **211**, 145–161 (2019). https://www.jstor.org/stable/48632501
19. Pavlopoulos, J., Konstantinidou, M., Marthot-Santaniello, I., Essler, H., Paparigopoulou, A.: Dating Greek papyri with text regression. In: Rogers, A., Boyd-Graber, J., Okazaki, N. (eds.) Proceedings of the 61st Annual Meeting of the Association for Computational Linguistics (Volume 1: Long Papers), pp. 10001–10013. Association for Computational Linguistics, Toronto, Canada (2023). https://doi.org/10.18653/v1/2023.acl-long.556, https://aclanthology.org/2023.acl-long.556
20. Pavlopoulos, J., et al.: Explaining the chronological attribution of Greek papyri images. In: Bifet, A., Lorena, A.C., Ribeiro, R.P., Gama, J., Abreu, P.H. (eds.) DS 2023. LNCS, vol. 14276, pp. 401–415. Springer, Cham (2023). https://doi.org/10.1007/978-3-031-45275-8_27
21. Seuret, M., et al.: ICDAR 2021 competition on historical document classification. In: Lladós, J., Lopresti, D., Uchida, S. (eds.) ICDAR 2021. LNCS, vol. 12824, pp. 618–634. Springer, Cham (2021). https://doi.org/10.1007/978-3-030-86337-1_41
22. Vandorpe, K.: The bilingual family archive of Dryton, his wife Apollonia and their daughter Senmouthis (P.Dryton). Koninklijke Vlaamse Academie van België voor Wetenschappen en Kunsten, Brussels (2002)
23. Wahlberg, F., Wilkinson, T., Brun, A.: Historical manuscript production date estimation using deep convolutional neural networks. In: 2016 15th International Conference on Frontiers in Handwriting Recognition (ICFHR), pp. 205–210 (2016). https://doi.org/10.1109/ICFHR.2016.0048

Automated Dating of Medieval Manuscripts with a New Dataset

Boraq Madi[1,2](✉), Nour Atamni[1](✉), Vasily Tsitrinovich[2],
Daria Vasyutinsky-Shapira[3], Jihad El-Sana[1], and Irina Rabaev[2]

[1] Ben-Gurion University of the Negev, Beer-Sheva, Israel
madibo@ac.sce.ac.il, {atamni,el-sana}@post.bgu.ac.il
[2] Shamoon College of Engineering, Beer-Sheva, Israel
irinar@ac.sce.ac.il
[3] Tel Aviv University, Tel Aviv, Israel
dariashap@tauex.tau.ac.il

Abstract. Automated manuscript dating is a long-awaited valuable tool for scholars in their research of historical documents. This study presents a new dataset of medieval Hebrew manuscripts annotated with dates. Our initial experiments focus on documents written in the Ashkenazi square script, allowing us to refine our methodologies in a manageable setting before addressing more complex script types. Also, to accurately reflect the script's historical evolution, we adopt a novel classification approach for time periods of varying lengths, which acknowledges the uneven development of the script over time. We perform extensive experimentation with a variety of deep-learning models and show that the regression approach is more appropriate for estimating the date of the manuscript compared to categorical classification.

Keywords: Historical document images · Automated dating · Historical dataset · Regression · Classification

1 Introduction

Automated manuscript dating holds potential applications across various domains. It is a valuable tool for aiding scholars in analyzing undated or debated works. Additionally, the integration of automatic manuscript dating can help researchers in historical linguistics, archeology, sociology, and also enhance the relevance of results in the field of information retrieval.

D. Vasyutinsky-Shapira—The participation of Dr. Daria Vasyutinsky Shapira was funded by the European Union (ERC, MiDRASH, Project No.@ 101071829). Views and opinions expressed are, however, those of the author only and do not necessarily reflect those of the European Union or the European Research Council Executive Agency. Neither the European Union nor the granting authority can be held responsible for them.

In conventional manuscript analysis, manual processing is the primary method of analysis, which requires researchers to possess specialized expertise in paleography, a field that focuses on studying historical writing systems. However, the growing digitization of historical documents over the last few decades necessitated the development of computational techniques to process and analyze digital collections. Automated dating methods can contribute to streamlining and improving the accuracy of dating processes, making scholarly research and information retrieval more efficient and effective.

Several studies have explored the task of automatic dating of historical manuscripts [17,30,36,42,44,50], including dedicated competitions [7,8,35]. However, the majority of the studies focused on Latin scripts. There are still a number of languages with significantly different writing characteristics that remain understudied. Particularly, to the best of our knowledge, no studies systematically explored automated dating of historical documents in Hebrew.

Our key contributions include:

- We present a new dataset of medieval documents annotated with date. In this study, we focus on manuscripts written in the Ashkenazi square script. Starting with this well-defined subset enables us to develop and refine our methodologies in a controlled, manageable environment before we expand our analysis to the lesser studied script types.
- To more accurately and logically reflect the script's evolution, we propose a classification of time periods that vary in length. This approach recognizes the unequal rates of change and development that the script underwent at different historical stages.
- We conducted extensive experiments using a variety of models and compared regression versus categorical classification approaches. We argue that the regression approach is better suited for the date estimation task. We also evaluated and compared the page-level and manuscript-level scenarios. Our experiments lay the groundwork for future research.

We believe that this research contributes not only to the field of medieval Hebrew manuscripts studies but also has broader implications for advancing automated date estimation techniques in manuscript analysis.

2 Related Work

Driven by the need to assist scholars and reduce manual effort, various methods for automated dating of historical manuscripts have been developed over the last decades. These methods range from statistical models to deep neural networks.

Earlier works leveraged traditional machine learning methods, including statistical models [16,40,44], Random Forests [37,39], SVM [5], SVR [6], and Self-Organizing Maps [18,19].

In recent years, deep learning-based approaches have gained prominence in almost all fields, including automated document analysis. Li et al. [22] proposed a CNN-based approach in enhancing OCR accuracy by leveraging estimated publication dates. Wahlberg et al. [45] focused on Swedish historical manuscripts, using a CNN-based GoogleNet architecture for feature extraction and SVR for dating, and showed the potential of CNNs to learn features for accurate manuscript dating. Sidorov [36] explored paleographic dating of Birch Bark Manuscripts, employing an elastic model of grapheme deformation and features learned by CNNs. Studer et al. [38] investigated the impact of ImageNet pre-training on historical document analysis tasks. The research assessed pre-training effects on character recognition, style classification, manuscript dating, semantic segmentation, and content-based retrieval, providing insights into the benefits of pre-training in diverse tasks. Hamid et al. [17] utilized CNNs for automatic feature learning from small patches. The study explored writing styles' matching as a basis for estimating the date of manuscript origin. Yu and Huangfu [50] employed a Long-Short Term Memory Network (LSTM) to capture non-linear relationships within character sequences for automated dating of ancient Chinese texts. Series of the ICFHR and ICDAR competition [7,8,35] provided a comprehensive overview of historical document classification tasks, including font group/script type, location, and date estimation. The competitions showed varying difficulty levels of these classification tasks. Adam et al. [1] introduced a hierarchical fusion of traditional and ResNet-extracted features for dating historical manuscripts in Arabic. Molina et al. [27] proposed a system for historical document retrieval for date estimation of document collections, using a ranking loss function, smooth-nDCG, to train a CNN for document ordination. Assael et al. [2] presented "Ithaca" – a deep neural network designed to restore, attribute, and date ancient Greek inscriptions. The papers of Paparrigopoulou et al. [30,31] addressed automated dating of Greek papyri images with several classifiers, both traditional machine learning (SVM, decision trees, Random Forests) and deep-learning (DenseNet, VGG, EfficientNet, ResNet). Pavlopoulosu et al. [32] trained several regression models to predict dates for Greek papyri and leverage a Convolutional Neural Network to provide precise date estimations for papyri with disputed or vaguely defined dates [33]. Tvalavadze et al. [42] utilized the CNN for automated dating of the works of Georgian poet Galaktion Tabidze.

Several studies have explored the automated analysis of Hebrew manuscripts. The early work of Wolf et al. [46] experimented with the script style classification on the Cairo Genizah collection. Prebor et al. [34] introduced a methodology to predict migration patterns and locations by combining script type with temporal data. The works of Vasyutinsky Shapira et al. [43] and Droby et al. [10,12] investigated the use of deep-learning models for classifying medieval Hebrew scripts using the VML-HP dataset. Feigenbaum-Golovin et al. [15] explored computational paleography for ancient Hebrew inscriptions and analyzed processing methods, including binarization, letter segmentation, automatic handwriting analysis, and writer identification. Droby et al. [11] compared hard and soft-labeling approaches for script classification. Naamneh et al. [28] focused on Aramaic incantation bowls, introducing a new dataset and proposing a Multi-Level-of-Detail architecture for improved accuracy of script similarity analysis.

3 Dataset Overview

Sfardata is the name of a database created by the Hebrew Paleography Project, under the direction of Malachi Beit-Arié, of blessed memory. The Hebrew Palaeography Project was established in 1965 with the goal of inventorying and studying all medieval Hebrew codices prior to 1540 which contain an explicit mention of their date or can be precisely dated. Consequently, the manuscripts were located in more than 250 public and private collections around the world, and their codicological and paleographical features were described. Eventually, the project described more than 3,300 manuscripts within the time span from the 9*th* century to 1540; some contain the mention of the place of copying.

In the early 1970s Malachi Beit-Arié initiated a pioneering effort to convert the documented data into a computerized form. Numerous changes were introduced following the development of computer technologies, and eventually, in the early 2010s, SfarData was converted to an online database. Today, it is hosted by the National Library of Israel.

Paleographically, manuscripts in the Sfardata are described according to their script mode (square, cursive, semi-cursive) and to their geographical type. The six geographical types are Oriental (Egypt, Palestine, Syria and Lebanon, Iraq, Iran, Uzbekistan and Bukhara, Eastern Turkey); Sephardic (the Iberian Peninsula, Provence and Languedoc, North Africa, and Sicily); Italian; Ashkenazi (France and England, the Holy Roman Empire, Central, and Eastern Europe), Byzantine (Greece, the Balkans, Western Asia Minor, and regions surrounding the Black Sea); and Yemenite.

Each manuscript is illustrated by two or more photographs. One of the photographs is often the page with the colophon (the scribe's note indicating the name of the scribe, the date, and often the place of copying). The colophon is often scribed in a different script type mode than the main text.

The compiled dataset contains 299 accurately labeled page images excerpted from 166 different manuscripts written in Ashkenazi square script. The manuscripts are dated from 1176 to 1533, and their distribution over the years is presented in Appendix, Fig. 4. In this study, we focused our initial experiments exclusively on documents written in the Ashkenazi square script. This decision was made for several reasons. Firstly, there is a unique evolution of each Hebrew script type across different periods, which precludes a standardized period classification applicable to all types. Consequently, it is essential to define classifications individually for each script type. Secondly, the Ashkenazi script, along with Sephardic and Italian scripts, are relatively well-studied [4,13,14,29]. Square scripts are more uniform than semi-cursive and cursive script modes, and dating the later requires profound paleographical training. Providing automatic solutions for this problem is both challenging and rewarding. Finally, by beginning with a smaller, well-defined subset-such as the Ashkenazi square script-we can develop and refine our methodologies in a controlled, manageable context before extending our analysis to more varied and lesser studied script types. For the Ashkenazi square script, our team's paleographer proposed a classification into four classes, as listed in Table 1. This classification does not adhere to equal-length time periods; rather, it is logically structured to reflect the evolution of this script.

Table 1. Classes for Ashkenazi square manuscripts dating

Class	Years Interval	Number of Manuscripts	Number of Pages
1	1176 – 1250	9	20
2	1250 – 1306	39	66
3	1306 – 1400	54	96
4	1400 – 1533	64	117

4   Models Overview

In this section, we provide an overview of the diverse range of models employed in our experiments, each showcasing unique advancements and adaptations to address specific challenges in computer vision tasks.

We begin with the Vision Transformer (ViT) [9], an architecture that directly applies transformer principles to sequences of image patches, outperforming traditional convolutional networks while requiring fewer computational resources. ViT's success has spurred the development of several derivative models, each aiming to enhance performance and efficiency in image classification tasks.

Data-Efficient Image Transformers (DeIT) [41] introduce resource-efficient image classification models, leveraging a teacher-student training strategy to achieve competitive accuracy with reduced data and computing resources. Similarly, BERT Pre-Training of Image Transformers (BEiT) [3] integrates masked image modeling inspired by BERT, surpassing supervised pre-training methods and excelling in downstream tasks like image classification and semantic segmentation.

Document Image Transformer (DiT) [21] extends the self-supervised learning objective of BEiT to document images, achieving state-of-the-art results in document image classification and analysis tasks, addressing the scarcity of labeled document data.

The Swin Transformer [24] introduces a hierarchical architecture with shifted windows for efficient self-attention computation, achieving impressive performance across various tasks with linear computational complexity relative to image size. Swin Transformer V2 [23] further improves training stability and resolution adaptability, setting new benchmarks in large-scale model performance.

ConvNeXT [25,47] modernizes ResNets with vision Transformer design principles, demonstrating remarkable performance in image classification, object detection, and semantic segmentation tasks.

Convolutional Vision Transformer (CvT) [48] integrates convolutions into the ViT architecture, enhancing performance and efficiency, particularly on ImageNet-1k tasks, while MobileViT [26] combines lightweight CNNs with ViTs' global processing capabilities, achieving superior performance suitable for mobile devices.

FocalNet [49] introduces a novel focal modulation mechanism, replacing self-attention in vision transformers, and achieving superior performance across various tasks with comparable computational costs.

Each of these models represents a unique approach to leveraging transformer architectures in computer vision, showcasing advancements in performance, efficiency, and adaptability to address the evolving challenges in image processing tasks.

5 Experiments

We compare the results of the regression and classification approaches. Our hypothesis focuses on how different training-test split strategies impact model performance, particularly in the context of data regression and data classification tasks. We aim to determine whether the "Page-Level Split" or the "Manuscript-Level Split" yields superior results. This hypothesis suggests that the way we divide the dataset for training and testing can significantly influence model performance, especially in tasks involving predicting dates. By examining data regression and data classification problems, we seek to understand the effects of splitting strategies on model performance across these specific tasks.

5.1 Dataset Preparation Schemes

The dataset comprises 299 pages from 166 manuscripts spanning a period from 1176 to 1533, each associated with a specific year.

Our data preparation strategy involves two primary splitting strategies: page-level split and manuscript-level split. The page-level split randomly divides the document pages into training and testing sets. This methodology promotes model generalization across individual pages by incorporating variability across manuscripts. Specifically, pages from the same manuscript are distributed across both training and testing sets, allowing the model to learn from a diverse range of page layouts, contents, and styles. This split method is common in tasks where pages from the same manuscript can be treated as independent samples.

The manuscript-level split allocates pages associated with manuscripts originating from the same years to either the training or testing set exclusively. This ensures that the model's ability to generalize to entirely new manuscripts is rigorously tested. By mimicking real-life scenarios where models predict the labels of manuscripts they have not encountered during training, this approach evaluates the model's robustness and adaptability to unseen data.

We adhere to a standard split of 70% for training and 30% for testing datasets (split at page and manuscript levels). Within the training phase, a validation dataset is derived from the 20% of the training data. Our model training and testing procedures involve the extraction of patches from document images, following the patch generation method detailed in [10]. These patches are uniformly extracted based on text scale, with an average of 5 lines per patch as shown in Appendix, Fig. 5.

For the manuscript-level split, 6700 patches are extracted for training and 930 test set. For the page-level split, we include 9800 patches for the training set and 2160 for testing. Both regression and classification strategies maintain this split, ensuring equitable comparisons across tasks.

5.2 Models Preparation

As was mentioned in Sect. 4, the experimented were conducted with a variety of architectures of Vision Transformers, encompassing a diverse array of models such as Vision Transformer (ViT) [9], ConvNeXt [25,47], DeIT [41], FocalNet [49], Convolutional vision Transformers (CvT) [48], Shifted Window Transformer (Swin) [24], SwinV2 [23], MobileViT [26], BERT for Image Transformers (BEiT) [3], and Document Image Transformer (DiT) [21]. We also explored various versions of these models, including base, small, and tiny variants, as shown in Tables 2 and 3. The primary distinction among these versions lies in the size and complexity of the models, as well as the number of parameters they contain.

To adapt these models for regression tasks, we utilized a straightforward modification by removing the last layer of the classifier and appending a tensor with dimensions of 1×1. This configuration enabled the model to predict the year as a continuous variable. For classification tasks, we modified the models by truncating the last layer and replacing it with an output layer comprising four classes corresponding to specific intervals. This transformation allowed the models to categorize the manuscripts into distinct temporal periods. By implementing these adjustments, we tailored the architectures to suit the respective regression and classification objectives.

5.3 Regression Study

For the regression task, we constructed models according to the methodologies outlined in Sect. 5.2. To ensure consistency across all models, we employed the Adam optimizer [20] throughout the training process. Each model was initialized from scratch and trained for a duration of 20 epochs, with a fixed batch size of 8. The learning rate was set to 0.002 for all models to facilitate stable and efficient training. During the training process, input images were partitioned into patches of size 256×256, allowing the models to capture detailed spatial information. Mean Squared Error (MSE) was utilized as the primary loss function to quantify the discrepancy between predicted and ground truth year values. This choice of loss function is well-suited for regression tasks, enabling the models to minimize the deviation from the true year values during training. Through this standardized training protocol, we aimed to ensure uniformity in training procedures across all models and a fair comparison of their performance in regression tasks.

We utilized the MSE metric to calculate the error rate (ER). This metric measures the average error between the predicted and ground truth year values across the test set. Tables 2 and 3 list the ER values from the experiments over the test set. Both tables present the performance of regression models with

Table 2. Performance of Regression Models with Page-Level Split

Model	ER	ER with 5% Trim	ER with 10% Trim	Median ER
BEiT$_{base}$	25.45	19.23	15.10	5.0
ConvNeXt$_{base}$	25.14	18.69	14.69	5.0
ConvNeXt$_{tiny}$	26.35	21.17	17.37	12.0
CvT$_{13}$	28.97	23.90	19.68	13.0
CvT$_{21}$	28.80	23.76	20.39	16.0
DeIT$_{base}$	25.44	19.26	15.01	4.0
DeIT$_{small}$	27.94	21.22	16.85	6.0
DeIT$_{tiny}$	29.74	23.71	19.17	8.0
DiT$_{base}$	45.31	40.35	36.49	37.0
FocalNet$_{base}$	25.47	18.61	14.21	**3.0**
FocalNet$_{small}$	**21.87**	**16.04**	**11.95**	4.0
FocalNet$_{tiny}$	26.18	19.30	15.03	4.0
MobileViT$_{base}$	31.87	27.88	24.64	23.0
MobileViT$_{small}$	45.76	40.62	37.23	44.0
MobileViT$_{tiny}$	53.82	49.28	45.50	50.0
Swin$_{base}$	27.32	21.44	17.01	6.0
Swin$_{small}$	27.34	20.21	15.52	6.0
Swin$_{tiny}$	31.02	24.09	19.77	11.0
Swinv2$_{base}$	25.29	18.88	14.33	3.0
Swinv2$_{tiny}$	25.13	18.45	14.30	5.0
ViT$_{base}$	27.18	20.33	15.97	4.0

two different data preparation strategies: page-level split and manuscript-level split. For the page-level split (Table 2), FocalNet$_{small}$ achieved the top results, with the lowest ER of 21.87. MobileViT$_{tiny}$ exhibits the highest error rate of 53.82, indicating its relatively poor performance compared to other models in this approach. On the other hand, when considering the manuscript-level split, CvT$_{21}$ achieved the best results with the lowest error rate of 52.87.

Though the page-level split showed much lower error rates, the model might not truly learn to generalize across different manuscripts. It might learn specific features of the manuscript because its pages appear in both sets. In order to ascertain which approach, either the page-level split or manuscript split, is more effective, we have devised a rigorous evaluation plan. Firstly, we constructed a "blind dataset" that includes 5 pages from the manuscripts that were not used in our dataset, with each page contributing 30 patches, a total of 150 patches. None of these patches have been previously employed in either of the splitting schemes, i.e., the samples used for this evaluation are entirely separated from those utilized in constructing the training/test splits.

Table 3. Performance of Regression Models with Manuscript-Level Split

Model	ER	ER with 5% Trim	ER with 10% Trim	Median ER
BEiT$_{base}$	61.99	55.97	51.51	53.0
ConvNeXt$_{base}$	56.25	51.79	48.15	50.0
ConvNeXt$_{tiny}$	57.35	52.88	48.97	48.0
CvT$_{13}$	55.27	50.73	46.96	47.5
CvT$_{21}$	**52.87**	**48.27**	**44.43**	**40.5**
DeIT$_{base}$	56.90	52.10	48.16	47.0
DeIT$_{small}$	53.55	48.50	44.56	44.0
DeIT$_{tiny}$	61.54	56.35	52.14	53.5
DiT$_{base}$	64.19	59.46	55.47	56.0
FocalNet$_{base}$	62.53	56.10	52.05	54.0
FocalNet$_{small}$	62.15	57.08	53.00	54.0
FocalNet$_{tiny}$	61.30	56.45	52.32	54.0
MobileViT$_{base}$	54.16	49.65	45.78	47.0
MobileViT$_{small}$	53.99	49.27	45.20	44.0
MobileViT$_{tiny}$	53.31	47.74	43.85	43.0
Swin$_{base}$	59.57	54.28	49.57	43.5
Swin$_{small}$	58.25	53.74	50.06	46.0
Swin$_{tiny}$	62.01	56.15	51.21	50.0
Swinv2$_{base}$	59.80	54.26	49.76	48.0
Swinv2$_{tiny}$	55.29	50.72	46.75	44.0
ViT$_{base}$	58.20	53.32	49.07	50.0

Figures 1.a and 1.b illustrate the disparity between the performance of each model on the test set from the split (depicted by blue columns) and the blind test (highlighted by the green column) for page-level and manuscript-level approach. Based on the analysis of Fig. 1, it becomes evident that the page-level approach exhibits considerable weakness, with an average difference of 50.45 between the performance on the test set and the blind test. The page-level approach falls significantly in real-life scenarios. On the other hand, the manuscript-level split approach showcases a lower average difference in performance among models, averaging 10.9. This highlights the contrast between the manuscript-level approach and the page-level approach, emphasizing the effectiveness of the former in achieving more controlled and stable results in real-life scenarios. Therefore, we conclude that the manuscript-level split approach is preferable to applications where the model needs to work with previously unseen documents.

The analysis of Fig. 1.b provides insightful observations regarding the performance of different models in the context of the manuscript-level approach. Notably, models encompassed by dotted red rectangles achieve a remarkable feat, exhibiting less than a 5-year difference between their performance on the test set and the blind test. This observation underscores their robustness in constructing stable and reliable features, indicative of their capability to adapt effectively to unseen data scenarios. Other models, not surrounded by the highlighted rectangles, display a disparity of more than 5 years between their performance on the test set and the blind test.

Our investigation into the impact of extreme differences (outliers) on manuscript-level split regression models' performance highlights the sensitivity of MSE to values with significant deviations from the norm. These outliers can substantially distort MSE assessments, falsely indicating poorer model performance even when models perform adequately on most of the data. To address this, we analyzed the differences between mean and median error values for various models on both test and blind test datasets as shown in Appendix, Table 6 and Fig. 6.

The average difference between the mean and median values on the test dataset is approximately 8.73, while for the blind test dataset, it is significantly higher at approximately 17.56. This indicates a higher presence of outliers in the blind test dataset, leading to larger discrepancies between mean and median errors. Consequently, while evaluating model performance, it is crucial to consider both mean and median values, particularly in the presence of outliers, to avoid misleading assessments based on MSE alone.

Most of the models have a lower median (black line) than mean (red dotted line), such as CvT_{21}, $BEiT_{base}$ as shown in Appendix, Fig. 6. This leads to a significant difference between the mean and median errors, indicating the presence of a few high-error instances of outliers that skew the mean.

To mitigate the potential impact of outliers that could skew the evaluation results, we utilized a trimmed mean approach and median criterion. Specifically, we applied trimming parameters of 5% and 10% to data points that exhibit extreme deviations from the dataset's tails. By employing this trimmed mean methodology, we aimed to obtain a more robust and reliable estimation of the model's performance, which is less susceptible to the influence of anomalous data points. This comprehensive evaluation framework allowed us to assess the regression models' performance accurately and objectively, providing insights into their efficacy in predicting manuscript pages years with minimal bias and distortion from outliers.

5.4 Classification Study

The classification task was carried out as follows. We followed the methodologies outlined in Sect. 5.2. The Adam optimizer [20] was utilized throughout the training process. Each model underwent initialization from scratch and was trained for 10 epochs, with a constant batch size of 8. A learning rate of 0.002 was chosen for all models. Similarly to the regression experiments, throughout the training phase, input images were segmented into patches of dimensions 256×256. Cross Entropy Loss was adopted as the primary loss function to assess the disparity between predicted and ground truth class labels. This choice of loss function is

Table 4. Performance of Classification Models with Page-Level Split

Model	Precision (%)	Recall (%)	Model	Precision (%)	Recall (%)
BEiT$_{base}$	77.91	77.71	FocalNet$_{small}$	81.34	80.51
ConvNeXt$_{base}$	83.27	81.68	FocalNet$_{tiny}$	81.94	80.35
ConvNeXt$_{tiny}$	81.95	74.59	MobileViT$_{base}$	76.76	73.95
CvT$_{13}$	84.13	76.36	MobileViT$_{small}$	81.64	77.53
CvT$_{21}$	83.73	81.13	MobileViT$_{tiny}$	76.31	74.55
DeIT$_{base}$	82.89	79.38	Swin$_{base}$	82.28	73.81
DeIT$_{small}$	81.38	79.02	Swin$_{small}$	80.74	73.44
DeIT$_{tiny}$	80.68	74.13	Swin$_{tiny}$	81.06	73.69
DiT$_{base}$	79.57	74.49	Swinv2$_{base}$	81.97	83.42
FocalNet$_{base}$	80.60	77.70	Swinv2$_{tiny}$	**84.19**	**84.49**
ViT$_{base}$	78.54	76.51			

particularly well-suited for classification tasks, facilitating the minimization of the difference between predicted and actual class distributions during training.

Utilizing the same dataset, we can directly compare the performance of models trained under different splitting strategies, aiming to identify the optimal approach for our classification task. For the evaluation, we utilized Precision and Recall metrics. The classification results are presented in Table 4 and Table 5.

Based on Table 4, the model with the best precision is Swinv2$tiny$, with a precision of 84.19%, and the model with the best recall is also Swinv2$tiny$, with a recall of 84.49%. Therefore, Swinv2$_{tiny}$ achieves the highest precision and recall among the models listed in the table. In Table 5, among the models listed in the table, ViT$base$ has the highest precision (66.51%), indicating its ability to correctly identify positive instances. However, Swinv2$tiny$ has the highest recall (62.82%).

As in the regression task, we use a 'blind test' set to compare different dataset-split approaches. Figures 2.a and 2.b depict the disparity between precision and recall of each model on the test set (represented by green columns) and the blind test scores (blue columns) for the page-level approach. Upon analyzing Fig. 2, it becomes evident that the page-level approach exhibits significant weaknesses, particularly in effectively addressing the blind test data. With an average difference of 43.39% and 37.44% for precision and recall, respectively, between the performance on the test set and the blind test, the page-level approach falls short in real-life scenarios. Nevertheless, specific models, such as DeIT$_{base}$ and BEiT$_{base}$-highlighted by dotted red rectangles in Fig. 2.a-demonstrate exceptional performance, achieving precision differences of less than 10 between their performance on the test set and the blind test. This observation underscores their robustness in constructing stable and reliable features, indicative of their ability to adapt effectively to unseen data scenarios. Similarly, Fig. 2.b showcases models like DeIT$_{base}$-also encompassed by dotted red rectangles-achieving

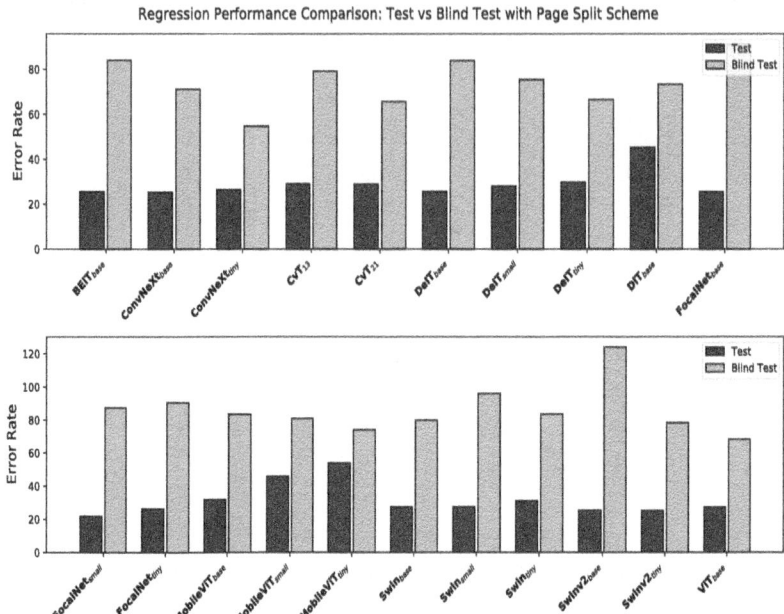

(a) Models error rate comparison over the test and blind test using page split scheme

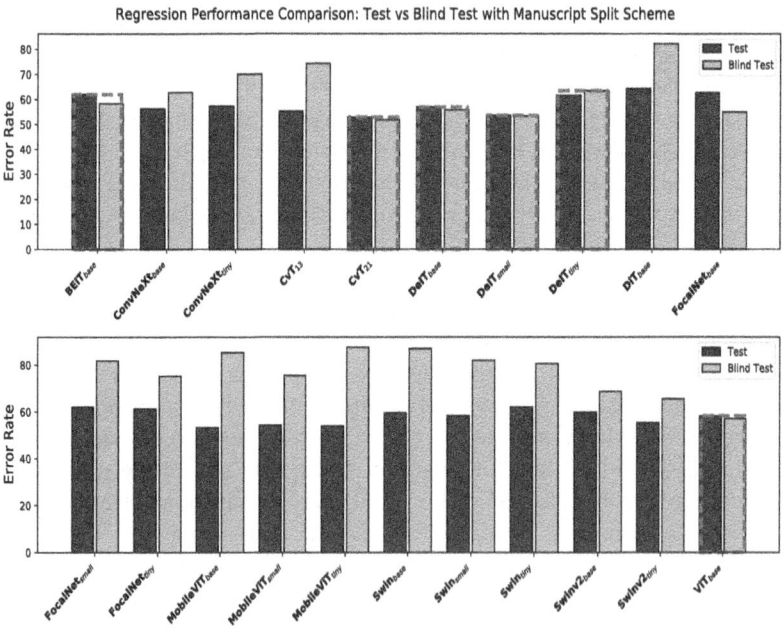

(b) Models error rate comparison over the test and blind test using manuscript split scheme

Fig. 1. Errors rate comparison over different models

Table 5. Performance of Classification Models with Manuscript-Level Split

Model	Precision (%)	Recall (%)	Model	Precision (%)	Recall (%)
$BEiT_{base}$	59.98	54.32	$FocalNet_{small}$	57.68	53.25
$ConvNeXt_{base}$	63.69	57.98	$FocalNet_{tiny}$	54.58	55.40
$ConvNeXt_{tiny}$	62.56	54.00	$MobileViT_{base}$	55.37	57.76
CvT_{13}	55.31	50.24	$MobileViT_{small}$	54.62	56.47
$\mathbf{CvT_{21}}$	64.62	60.45	$MobileViT_{tiny}$	51.12	49.81
$DeIT_{base}$	52.31	53.20	$Swin_{base}$	61.72	59.81
$DeIT_{small}$	52.78	44.83	$Swin_{small}$	65.17	59.16
$DeIT_{tiny}$	56.53	52.75	$Swin_{tiny}$	61.57	55.61
DiT_{base}	55.86	51.38	$Swinv2_{base}$	63.74	62.17
$FocalNet_{base}$	54.87	50.89	$Swinv2_{tiny}$	62.46	62.82
ViT_{base}	66.51	56.90			

recall differences of less than 10% between their performance on the test set and the blind test. Overall, while models trained on a page-level approach exhibit adaptation with blind datasets, $DeIT_{base}$ stands out as an exception, achieving consistently good and closely aligned results in the blind dataset.

Figures 3.a and 3.b reveal a low disparity between precision and recall in model performance across both the test set and blind test data, with a difference of only 10% between their performance on the test set and the blind test. These models are highlighted by dotted red rectangles, emphasizing their notable consistency. The minimal differences observed in both precision and recall averages, with values of 9.14% and 10.70%, respectively, suggest a promising level of consistency in model performance across both the test set and blind test data. This indicates that, on average, the models maintain stable and reliable performance when confronted with previously unseen data scenarios. Such minimal disparities indicate that the models effectively generalize their learned patterns to new instances, showcasing a robust ability to adapt to novel data environments.

In conclusion, while both approaches have their merits, the manuscript-level approach outperforms the page-level approach in terms of stability and adaptability, making it the preferred choice for real-world applications requiring robust performance across diverse scenarios.

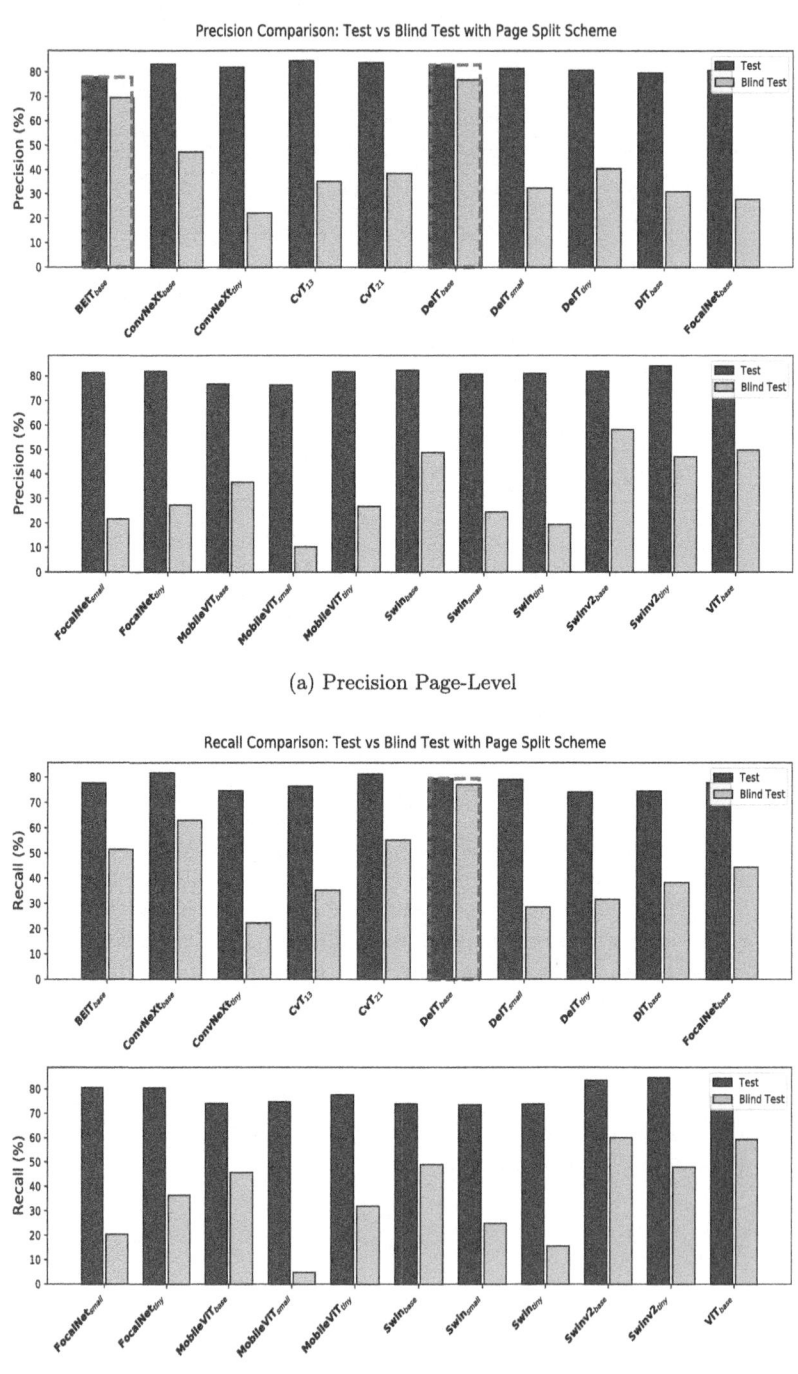

(a) Precision Page-Level

(b) Recall Page-Level

Fig. 2. Precision and Recall values for different models, page-level split

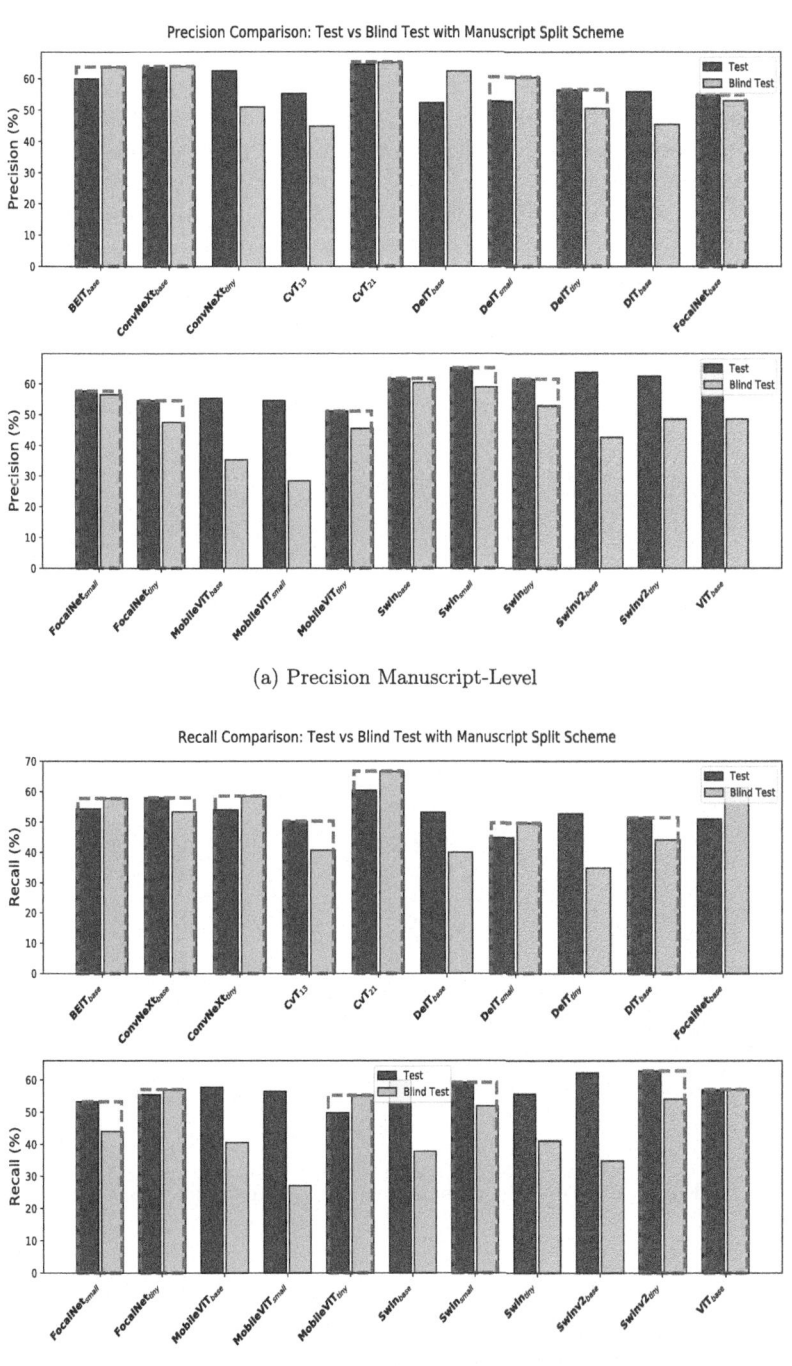

(a) Precision Manuscript-Level

(b) Recall Manuscript-Level

Fig. 3. Precision and Recall values for different models, manuscript-level split

6 Conclusions

In this study, we have introduced a new dataset of medieval Hebrew manuscripts. Our initial experiments focused on manuscripts written in the Ashkenazi square script. We implemented a classification system for time periods that vary in length to reflect the nuanced evolution of the script over different historical stages. We compared the regression and classification approaches with various deep-learning models for the date estimation task. We also compared and analyzed different dataset split approaches. We argue in favor of manuscript-level split. In addition, our findings confirm that regression approaches are superior for the task of date estimation when compared to categorical classifications. Currently, we are extending our dataset to include additional script styles and modes to perform future experiments on a more diverse set of manuscripts.

A Appendix

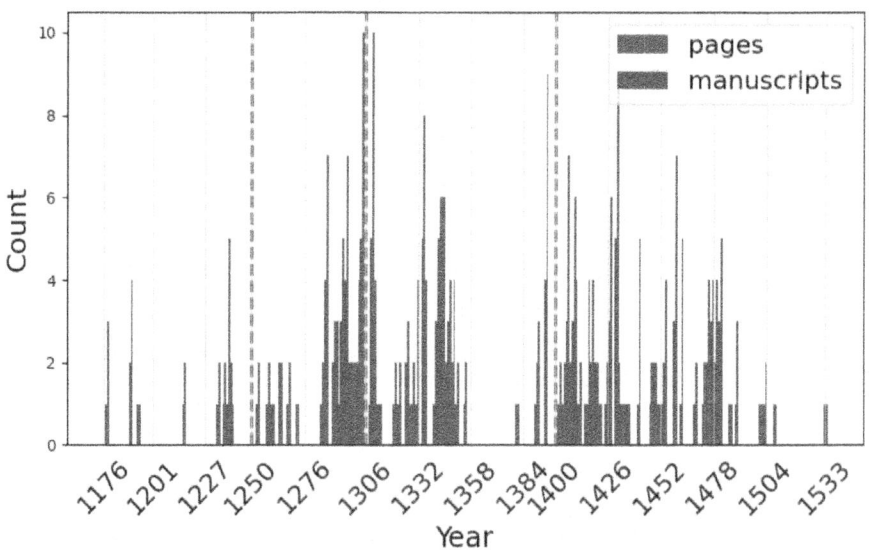

Fig. 4. Distribution of manuscripts/pages over the years

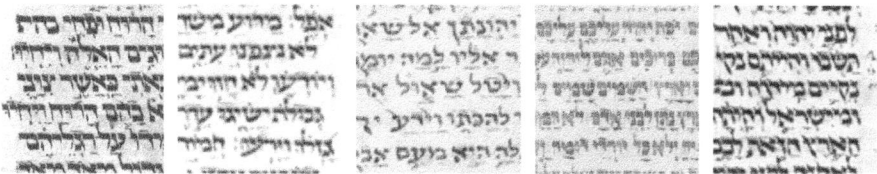

Fig. 5. Examples of the extracted patches. The patches are extracted with an average of 5 lines per patch.

Table 6. Mean and Median Errors for Test and Blind Test Sets

Model	Test		Blind Test	
	Mean	Median	Mean	Median
BEiT$_{base}$	61.99	53	58.29	33
ConvNeXt$_{base}$	56.25	50	62.77	49
ConvNeXt$_{tiny}$	57.35	48	70.02	65
CvT$_{13}$	55.27	47.5	74.34	79
CvT$_{21}$	52.87	40.5	51.8	26
DeIT$_{base}$	56.90	47	55.79	37.5
DeIT$_{small}$	53.55	44	53.29	35
DeIT$_{tiny}$	61.54	53.5	63.37	42
DiT$_{base}$	64.19	56	82.05	99
FocalNet$_{base}$	62.53	54	54.81	35
FocalNet$_{small}$	62.15	54	81.76	36
FocalNet$_{tiny}$	61.30	54	75.21	30
MobileViT$_{base}$	54.16	47	75.41	101.5
MobileViT$_{small}$	53.99	44	87.51	91
MobileViT$_{tiny}$	53.31	43	85.23	105
Swin$_{base}$	59.57	43.5	86.85	71
Swin$_{small}$	58.25	46	81.85	81
Swin$_{tiny}$	62.01	50	80.57	75
Swinv2$_{base}$	59.80	48	68.71	44
Swinv2$_{tiny}$	55.29	44	65.47	49
ViT$_{base}$	58.21	50	57.06	39

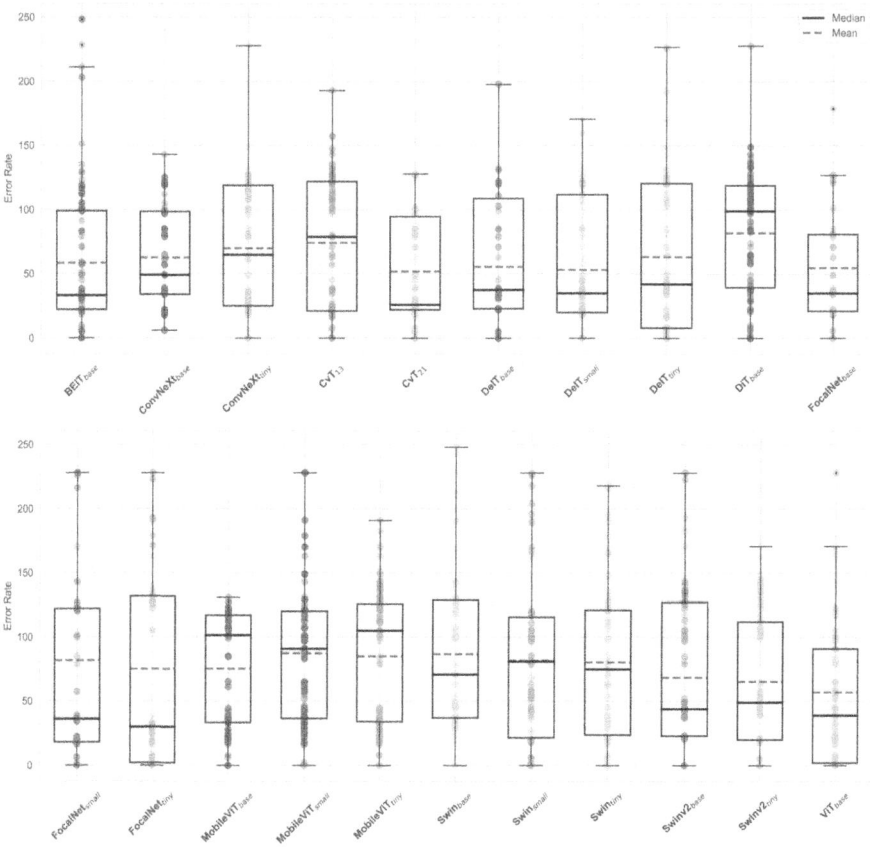

Fig. 6. The difference between the mean and median values on the blind test for each model

References

1. Adam, K., Al-Maadeed, S., Akbari, Y.: Hierarchical fusion using subsets of multi-features for historical Arabic manuscript dating. J. Imaging **8**(3), 60 (2022)
2. Assael, Y., et al.: Restoring and attributing ancient texts using deep neural networks. Nature **603**(7900), 280–283 (2022)
3. Bao, H., Dong, L., Piao, S., Wei, F.: BEiT: BERT pre-training of image transformers. In: International Conference on Learning Representations (2022). https://openreview.net/forum?id=p-BhZSz59o4
4. Beit-Arié, M., Engel, E.: Specimens of Mediaeval Hebrew scripts, vol. 3. Israel Academy of Sciences and Humanities (2017)
5. Boldsen, S., Paggio, P.: Automatic dating of medieval charters from Denmark. In: CEUR Workshop Proceeding (2019)
6. Christlein, V., Gropp, M., Maier, A.: Automatic dating of historical documents. Kodikologie und Paläographie im digitalen Zeitalter **4**, 151–164 (2017)

7. Cloppet, F., Eglin, V., Helias-Baron, M., Kieu, C., Vincent, N., Stutzmann, D.: Icdar2017 competition on the classification of medieval handwritings in Latin script. In: 2017 14th IAPR International Conference on Document Analysis and Recognition (ICDAR), vol. 1, pp. 1371–1376. IEEE (2017)
8. Cloppet, F., Eglin, V., Stutzmann, D., Vincent, N., et al.: ICFHR2016 competition on the classification of medieval handwritings in Latin script. In: 2016 15th International Conference on Frontiers in Handwriting Recognition (ICFHR), pp. 590–595. IEEE (2016)
9. Dosovitskiy, A., et al.: An image is worth 16x16 words: transformers for image recognition at scale. In: International Conference on Learning Representations (2021)
10. Droby, A., Kurar Barakat, B., Vasyutinsky Shapira, D., Rabaev, I., El-Sana, J.: VML-HP: Hebrew paleography dataset. In: Lladós, J., Lopresti, D., Uchida, S. (eds.) ICDAR 2021, Part IV 16. LNCS, vol. 12824, pp. 205–220. Springer, Cham (2021). https://doi.org/10.1007/978-3-030-86337-1_14
11. Droby, A., Rabaev, I., Shapira, D.V., Kurar Barakat, B., El-Sana, J.: Digital Hebrew paleography: script types and modes. J. Imaging **8**(5) (2022). https://doi.org/10.3390/jimaging8050143
12. Droby, A., Shapira, D.V., Rabaev, I., Barakat, B.K., El-Sana, J.: Hard and soft labeling for hebrew paleography: a case study. In: Uchida, S., Barney, E., Eglin, V. (eds.) DAS 2022. LNCS, vol. 13237, pp. 492–506. Springer, Cham (2022). https://doi.org/10.1007/978-3-031-06555-2_33
13. Engel, E.: Calamus or Chisel: On the History of the Ashkenazic Script, pp. 183 – 197. Brill, Leiden, The Netherlands (2010). https://doi.org/10.1163/ej.9789004179547.i-398.39
14. Engel, E.: Between France and Germany: gothic characteristics in Ashkenazi script. Nicholas de Lange and Judith Olszowy-Schlanger, Manuscrits hébreux et arabes: Mélanges en l'honneur de Colette Sirat, pp. 197–219 (2014)
15. Faigenbaum-Golovin, S., Shaus, A., Sober, B.: Computational handwriting analysis of ancient Hebrew inscriptions - a survey. IEEE BITS Inf. Theory Mag. **2**(1), 90–101 (2022). https://doi.org/10.1109/MBITS.2022.3197559
16. Feuerverger, A., Hall, P., Tilahun, G., Gervers, M.: Using statistical smoothing to date medieval manuscripts. In: Beyond Parametrics in Interdisciplinary Research: Festschrift in Honor of Professor Pranab K. Sen, vol. 1, pp. 321–332. Institute of Mathematical Statistics (2008)
17. Hamid, A., Bibi, M., Moetesum, M., Siddiqi, I.: Deep learning based approach for historical manuscript dating. In: 2019 International Conference on Document Analysis and Recognition (ICDAR), pp. 967–972 (2019). https://doi.org/10.1109/ICDAR.2019.00159
18. He, S., Samara, P., Burgers, J., Schomaker, L.: Discovering visual element evolutions for historical document dating. In: 2016 15th International Conference on Frontiers in Handwriting Recognition (ICFHR), pp. 7–12 (2016). https://doi.org/10.1109/ICFHR.2016.0015
19. He, S., Samara, P., Burgers, J., Schomaker, L.: Historical manuscript dating based on temporal pattern codebook. Comput. Vis. Image Underst. **152**, 167–175 (2016). https://doi.org/10.1016/j.cviu.2016.08.008
20. Kingma, D.P., Ba, J.: Adam: a method for stochastic optimization. arXiv preprint arXiv:1412.6980 (2014)
21. Li, J., Xu, Y., Lv, T., Cui, L., Zhang, C., Wei, F.: DIT: self-supervised pre-training for document image transformer (2022)

22. Li, Y., Genzel, D., Fujii, Y., Popat, A.C.: Publication date estimation for printed historical documents using convolutional neural networks. In: Proceedings of the 3rd International Workshop on Historical Document Imaging and Processing, pp. 99–106 (2015)
23. Liu, Z., et al.: Swin transformer v2: scaling up capacity and resolution. In: International Conference on Computer Vision and Pattern Recognition (CVPR) (2022)
24. Liu, Z., et al.: dosovitskiy2021an: Hierarchical vision transformer using shifted windows. In: Proceedings of the IEEE/CVF International Conference on Computer Vision (ICCV) (2021)
25. Liu, Z., Mao, H., Wu, C.Y., Feichtenhofer, C., Darrell, T., Xie, S.: A convnet for the 2020s. In: Proceedings of the IEEE/CVF Conference on Computer Vision and Pattern Recognition (CVPR) (2022)
26. Mehta, S., Rastegari, M.: MobileViT: light-weight, general-purpose, and mobile-friendly vision transformer. In: International Conference on Learning Representations (2022)
27. Molina, A., Gomez, L., Ramos Terrades, O., Lladós, J.: A generic image retrieval method for date estimation of historical document collections. In: Uchida, S., Barney, E., Eglin, V. (eds.) DAS 2022. LNCS, vol. 13237, pp. 583–597. Springer, Cham (2022). https://doi.org/10.1007/978-3-031-06555-2_39
28. Naamneh, S., et al.: Classifying the scripts of Aramaic incantation bowls. In: Proceedings of the 7th International Workshop on Historical Document Imaging and Processing, pp. 55–60. Association for Computing Machinery, New York, NY, USA (2023). https://doi.org/10.1145/3604951.3605510
29. Olszowy-Schlanger, J.: The early developments of Hebrew scripts in North-Western Europe. Gazette du livre médiéval **63**(1), 1–19 (2017)
30. Paparrigopoulou, A., Kougia, V., Konstantinidou, M., et al.: Greek literary papyri dating benchmark. Preprint 2272076 (2023). https://doi.org/10.21203/rs.3.rs-2272076/v2
31. Paparrigopoulou, A., Pavlopoulos, J., Konstantinidou, M.: Dating Greek papyri images with machine learning (2022). https://doi.org/10.21203/rs.3.rs-2272076/v1
32. Pavlopoulos, J., Konstantinidou, M., Marthot-Santaniello, I., Essler, H., Paparigopoulou, A.: Dating Greek papyri with text regression. In: Rogers, A., Boyd-Graber, J., Okazaki, N. (eds.) Proceedings of the 61st Annual Meeting of the Association for Computational Linguistics (Volume 1: Long Papers), pp. 10001–10013. Association for Computational Linguistics, Toronto, Canada (2023). https://doi.org/10.18653/v1/2023.acl-long.556
33. Pavlopoulos, J., et al.: Explaining the Chronological Attribution of Greek Papyri Images, pp. 401–415 (2023). https://doi.org/10.1007/978-3-031-45275-8_27
34. Prebor, G., Zhitomirsky-Geffet, M., Miller, Y.: A new analytic framework for prediction of migration patterns and locations of historical manuscripts based on their script types. Digit. Scholarsh. Human. **35**(2), 441–458 (06 2019). https://doi.org/10.1093/llc/fqz038
35. Seuret, M., et al.: ICDAR 2021 competition on historical document classification. In: Lladós, J., Lopresti, D., Uchida, S. (eds.) ICDAR 2021. LNCS, vol. 12824, pp. 618–634. Springer, Cham (2021). https://doi.org/10.1007/978-3-030-86337-1_41
36. Sidorov, K.: Paleographic dating of birch bark manuscripts. In: GraphiCon 2017, pp. 162–168 (2017)
37. Soumya, A., Kumar, G.H.: Classification of ancient epigraphs into different periods using random forests. In: 2014 Fifth International Conference on Signal and Image Processing, pp. 171–178 (2014). https://doi.org/10.1109/ICSIP.2014.33

38. Studer, L., et al.: A comprehensive study of imagenet pre-training for historical document image analysis. In: 2019 International Conference on Document Analysis and Recognition (ICDAR), pp. 720–725 (2019). https://doi.org/10.1109/ICDAR.2019.00120
39. Tagami, D., Satlow, M.: Machine learning techniques for analyzing inscriptions from israel. DHQ: Digit. Human. Q. **17**(2) (2023)
40. Tilahun, G., Feuerverger, A., Gervers, M.: Dating medieval English charters. Ann. Appl. Stat. **6**(4), 1615–1640 (2012). https://doi.org/10.1214/12-AOAS566
41. Touvron, H., Cord, M., Douze, M., Massa, F., Sablayrolles, A., Jegou, H.: Training data-efficient image transformers distillation through attention. In: International Conference on Machine Learning, vol. 139, pp. 10347–10357 (2021)
42. Tvalavadze, T., Gigashvili, K., Mania, E., Iavich, M.: Automated dating of Galaktion Tabidze's handwritten texts. In: Hu, Z., Dychka, I., He, M. (eds.) ICCSEEA 2023. LNDE and CT, vol. 181, pp. 260–268. Springer, Cham (2023). https://doi.org/10.1007/978-3-031-36118-0_23
43. Vasyutinsky Shapira, D., Rabaev, I., Droby, A., Barakat, B.K., El-Sana, J.: Is a deep learning algorithm effective for the classification of medieval Hebrew scripts? Studies in Digital History and Hermeneutics, p. 349 (2022). https://doi.org/10.1515/9783110744828-016
44. Wahlberg, F., Mårtensson, L., Brun, A.: Large scale continuous dating of medieval scribes using a combined image and language model. In: 2016 12th IAPR Workshop on Document Analysis Systems (DAS), pp. 48–53 (2016). https://doi.org/10.1109/DAS.2016.71
45. Wahlberg, F., Wilkinson, T., Brun, A.: Historical manuscript production date estimation using deep convolutional neural networks. In: 2016 15th International Conference on Frontiers in Handwriting Recognition (ICFHR), pp. 205–210 (2016). https://doi.org/10.1109/ICFHR.2016.0048
46. Wolf, L., Dershowitz, N., Potikha, L., German, T., Shweka, R., Choueka, Y.: Automatic palaeographic exploration of genizah manuscripts. In: Kodikologie und Paläographie im digitalen Zeitalter 2 - Codicology and Palaeography in the Digital Age 2, pp. 157–179. Books on Demand (BoD), Norderstedt (2011)
47. Woo, S., et al.: Convnext v2: co-designing and scaling convnets with masked autoencoders. In: Proceedings of the IEEE/CVF Conference on Computer Vision and Pattern Recognition, pp. 16133–16142 (2023)
48. Wu, H., et al.: CVT: introducing convolutions to vision transformers. arXiv preprint arXiv:2103.15808 (2021)
49. Yang, J., Li, C., Dai, X., Gao, J.: Focal modulation networks (2022)
50. Yu, X., Huangfu, W.: A machine learning model for the dating of ancient Chinese texts. In: 2019 International Conference on Asian Language Processing (IALP), pp. 115–120 (2019). https://doi.org/10.1109/IALP48816.2019.9037653

Image-to-Image Translation Approach for Page Layout Analysis and Artificial Generation of Historical Manuscripts

Chahan Vidal-Gorène(✉)🆔 and Jean-Baptiste Camps🆔

École nationale des chartes, Université Paris, Sciences & Lettres,
65 rue de Richelieu, 75002 Paris, France
{chahan.vidal-gorene,jean-baptiste.camps}@chartes.psl.eu

Abstract. Document layout analysis is essential in Optical Character Recognition (OCR) and Handwritten Text Recognition (HTR), especially for historical and low-resource scripts. This study explores a novel data augmentation technique using Generative Adversarial Networks (GANs) to generate realistic document layouts from semantic masks, enhancing layout analysis without increasing human annotation effort.

Our lightweight pipeline, tested on historical manuscripts (Latin, Arabic, Armenian, Hebrew), newspapers, and complex document layouts, shows that GAN-generated layouts are convincing and difficult to distinguish from real ones, even for paleographers. This method significantly boosts data augmentation, yielding a 3% point improvement in layout analysis metrics (precision, recall, mAP), and a 12 point increase in precision and recall for damaged documents. Additionally, masks with character information enhance image quality, boosting text recognition performance.

Keywords: GAN · Layout Analysis · Semantic Classification · Data Augmentation · Handwritten Text Recognition

1 Introduction

Layout analysis is a core task in the analysis and understanding of documents, that involves the detection and annotation of physical areas on the source material; its numerous applications range from document categorisation to text recognition, and is generally intended as a pre-processing step [4], that will affect the results and accuracy of all ulterior treatments. Many approaches exist, the most common being to separate the detection of layout areas (e.g., text columns, marginal notes, illustrations, figures, etc.) and of text lines: the page is first divided into regions, then the lines are detected within the relevant regions.

These steps are largely covered by state-of-the-art systems [9,10], in particular to detect baselines and propose a semantic classification of text regions (e.g. marginal note, title, etc.). Many datasets offer a two-level annotation of their contents [25], and the use of an ontology like SegmOnto [12] allows their pooling in order to enlarge the training datasets and lead to more versatile models.

In the case of historical documents, the task to overcome is more complex due to the very wide variety of layouts, physical support and page preparation, size of scripts, handwritten traditions, irregularities, damage to documents (e.g. worm holes, burnt manuscripts,...) or even scan quality. The creation of datasets for complex documents comes up regularly, betting for example on the variety [9], or on a family of manuscripts [19]. Despite the availability of increasingly versatile layout analysis models that enable document pre-annotation within mainstream platforms [18,20,37], any historical document processing project must even today consider a massive annotation of layouts in order to overcome a given corpus with a specialised model. Document annotation, even semi-automated, remains a time-consuming task, and incompatibilities between datasets (see Sect. 4) delay the arrival of very versatile models for historical documents. We therefore propose a new method of data augmentation, for the generation of artificial pages with complex layouts emulating historical documents, and evaluate it on several datasets and use cases. In particular, we envision the case of historical manuscripts, such as those damaged by fire.

2 Related Work

Recent efforts in layout analysis are integrating the detection of text regions and text lines into a singular task, primarily through the use of transformers [5,24,41], which achieve scores comparable to traditional state-of-the-art methods. However, these transformer-based approaches often require a large volume of data. Alternative methods include simple CRNN layers [19], effective in clear cases but struggling with closely situated regions of the same type. In contrast, U-net models [14,22,32,37] have led recent competitions [10] with less data dependency. Object detection strategies [8] also show promise, frequently surpassing other methods by leveraging extensively pre-trained models [24].

Generative adversarial networks (GANs) are increasingly employed for qualitative dataset augmentation [33] and have begun to influence document analysis with the integration of transformers in scientific document processing [1,30]. These are trained on extensive datasets such as scientific articles [28,42] and have been used to generate both printed [6] and handwritten layouts [22] on varied corpora, in particular modern handwriting [31]. Diffusion models have also been explored for generating scientific documents without text constraints, but only produce low-resolution results [35]. Despite the challenges posed by historical documents due to high noise levels, transformers are becoming feasible for these as data scarcity issues are addressed, improving HTR line performance [3]. GANs, with or without transformers, have been primarily evaluated for reducing the Character Error Rate (CER) of HTR models by generating fake text lines [11,36,40], achieving notable success in both Latin and non-Latin scripts [33]. However, their application for layout analysis in historical documents has been very limited.

While no definitive metric exists to assess the quality of GANs for text image generation (see Subsect. 5.2), these technologies are capable of producing qualitatively convincing layouts and text lines, except for figures and graphs. Most methods rely on a constrained approach where GANs generate layouts from semantic region coordinates provided in an XML file. This results in well-structured outputs, such as printed scientific articles or handwritten tables. A style transfer approach has also been tested on historical documents [39], applying a handwritten style to a printed template. While the clarity of template text boxes leads to an effective imitation, it fails to represent the deterioration found in historical documents. The same applies to complex documents like newspapers.

3 Proposed Method

We propose a method of artificial page generation, that accounts both for the constraint of emulating existing and relatively specific historical layouts, and being able of generating entirely new artificial pages, without over-specifying their composition. For this, we propose a method that relies on constrained image-to-image translation, generating artificial pages from input layout masks of regions and baselines.

The objective of this approach is to propose a framework constrained to GANs, utilizing explicit minimalist semantic information (regions, lines, or characters). A historical handwritten document style mapping is then applied to these semantic areas, allowing the GAN to focus solely on reproducing the writing aesthetic and the background. Our approach reuses the implementation provided by Pix2Pix [17]. The GAN establishes a relationship between image pixels and generated masks for object layouts. Generating false layout masks and transforming them into fake pages with known coordinates, inspired by the creation of maps or building facades [17], is less time-consuming than annotating documents. For historical documents, Pix2Pix has already been used to synthesis data for palimpsests reconstruction [21].

For each dataset, given sufficient data (see Sect. 4), we generate semantic masks for text regions and baselines, as illustrated in Fig. 1. This process allows for a dataset of masks equivalent to the image dataset. We are currently exploring both text region semantic masks and line masks. We also apply the same method at the character level to evaluate its transfer to a more qualitative task.

At the image generation stage, a semantic mask generator creates random masks, saving object coordinates in an XML file. Our GAN, pre-trained on a document type, then generates a synthetic image. This results in an XML of semantically tagged coordinates and an image (see Fig. 1). In details, we give to Pix2Pix random patches of 512 × 512px, cropped from a rescale image of 2048px height, with the Unet-256 model Pix2Pix and 300 hidden-size for the Generator. The style is unconstrained; only the semantic spatial information's position is dictated by the generator.

After data creation, we train a CRNN for layout analysis to verify the potential of this method to create gains in the accuracy of layout analysis tasks.

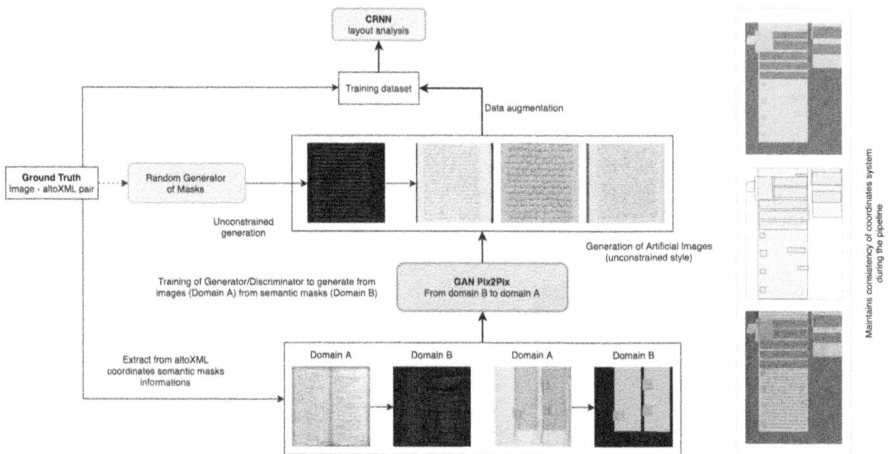

Fig. 1. Training pipeline using Pix2Pix at Region level and Baseline level

For generating images with explicit textual information, data augmentation occurs dynamically throughout the recognizer's training. This approach involves pre-training the recognizer for several epochs, then predicting on its training data to achieve a Character Error Rate (CER) under 1%. This prediction helps obtain character coordinates for generating varied random mask compositions.

The resulting images, constituting up to 25% of the training data, provide additional data augmentation without stylistic constraints, though mask dimensions may suggest a style to the GANs (see Fig. 2).

Image-to-image translation has demonstrated robust performance across many real-world applications. Despite its age compared to newer GAN models, supervised methods generally outperform unsupervised ones in image-to-image translation, even though advanced architectures like StyleGAN combined with contrastive learning techniques also show promising results [16,27]. These approaches, however, are not evaluated in our experiments.

4 Datasets

We conducted our main experiments on layout on three types of data:

1. **Manuscripts in Latin scripts**: CREMMA Medieval [29] with manuscripts in Old French and Latin, as well as cBAD 2017 Simple & Complex Track [9], that contains samples from documents written between 1470 and 1930, and coming from nine different European archives in Belgium, England, Finland, Germany, Italy and Switzerland [13].
2. **Manuscripts in non-Latin scripts**: RASAM for Arabic Maghribi [38], BADAM for Arabic scripts [19], and BiblIA for medieval Hebrew [34].
3. **Complex printed documents in Latin types** from the NewsEye READ dataset of contemporary French newspapers [23].

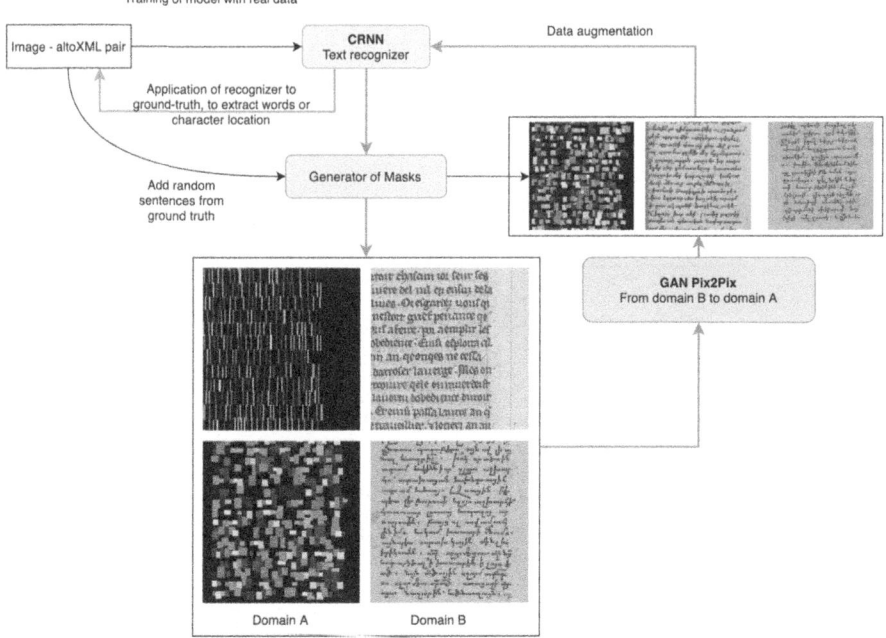

Fig. 2. Training pipeline using Pix2Pix at the character level

Experiments were conducted separately on each dataset to maintain control over the style generated and accommodate the varying levels of annotation detail provided for baselines and text regions.

The CREMMA Medieval and RASAM datasets served as our primary evaluation sources, with the CREMMA Medieval dataset offering comprehensive annotations using the SegmOnto ontology. Conversely, the BADAM, BiblIA, and NewsEye datasets were employed to test the viability of our approach on non-Latin scripts and printed documents with complex layouts due to their partial annotations (incomplete regions and lines).

The cBAD dataset, while included, presented challenges due to imprecise and insufficiently detailed annotations, rendering it less effective for our specific tasks. A detailed summary of each dataset's characteristics is provided in Table 1.

Complementary experiments on character and layout analysis tasks were conducted using the CREMMA Medieval, textscbadam, textscrasam, textsccbad Simple Track, and NewsEye datasets, chosen for their representation of diverse tasks. Additionally, a specialized Armenian dataset, not available in open access, was utilized to quantitatively evaluate the results and highlight specific paleographic patterns produced by Pix2Pix.

Table 1. Composition of datasets

Dataset	Baselines	Regions	Annotation	Ontology	Pages
Manuscripts					
Latin scripts					
CREMMA Medieval	✓	✓	full	SegmOnto	600
cBAD 2017 Simple	✓	✓	full	cBAD	703
cBAD 2017 Complex	✓	✓	full	cBAD	1060
Non-Latin scripts					
RASAM (Arabic Maghribi)	✓	✓	full	custom	300
BADAM (Arabic, mixed)	✓	-	partial	none	400
BIBLIA (Hebrew)	✓	✓	partial	none	132
Printed documents					
Latin types					
NewsEye	-	✓	full	custom	630

5 Results

Example of generations of artificial pages output by our model are presented in Table 2 for Lines and TextRegions generation, as well as in Fig. 3 and Fig. 4 for Lines and Characters generation. In our experiments, neither the diversity of objects in the dataset, nor the number of training samples appeared to significantly impact the model's ability to generate convincing images (e.g. while the CREMMA-Medieval dataset contains several pages of each source manuscript, the cBAD complex is composed only of isolated pages from quite diverse documents, without any obvious effect on the model generations).

5.1 Qualitative Assessment

The results across various datasets display a significant level of verisimilitude, particularly when evaluated holistically without focusing intensely on the textual content or the minutiae of the generated images. At the layout level, the models reproduce structural information that closely mirrors actual documents.

In the CREMMA dataset, for instance, the generated layouts accurately reflect the typical two or three-column configurations found in medieval French and Latin manuscripts. Additionally, the models are able to capture and emulate to a point finer details such as secondary decorations including initial letters, pen-flourishing extending into margins, and detached verse initials situated in their own columns. Major decorative elements like illuminations and painted miniatures are also replicated (Table 2). The pages generated solely from baseline data interestingly allow the model to infer semantic units such as columns and adequately fill spaces designated for decorations, without explicit semantic information in the mask regarding the presence of such a type of zone. These

Table 2. Output artificial images, generated by our model, based on the training on various datasets. The input mask is presented on the left, and the resulting generation on the right. The examples displayed here are sometimes strongly realistic on the basis of a general expert assessment

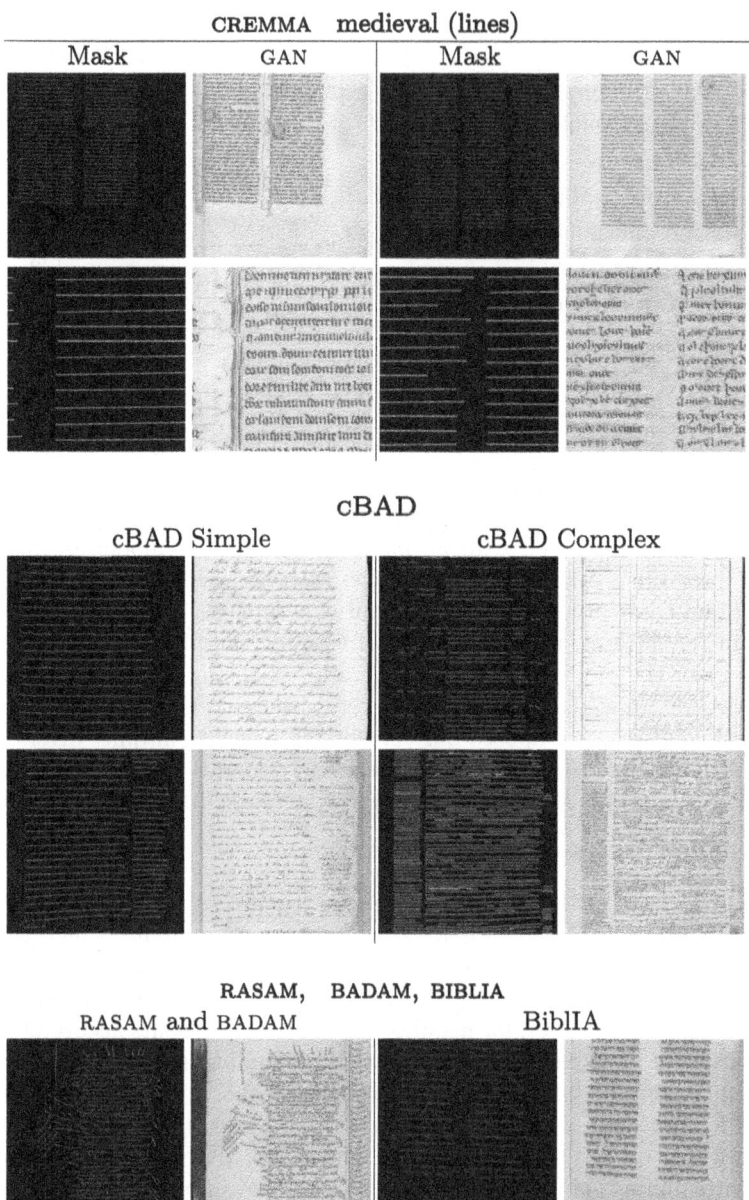

baseline-derived layouts present a more irregular, authentic appearance, which is characteristic of such manuscripts, as opposed to those generated from semantic regions that tend to appear overly regular due to the geometric shapes of the masks used. Furthermore, the model demonstrates the ability to emulate different script styles and hands, offering variations of the Gothic *Textualis* script with varying degrees of roundness and formality. However, the models still struggle to produce convincing illuminations or coherent text, often repeating the same words multiple times within a line, due to the lack of an integrated language model. This issue is particularly noticeable when more empty pages with minimal text are provided.

Similar patterns are observed in the cBAD dataset (Table 2), where the model successfully interprets elements like table layouts. The scripts' cursiveness is well-rendered, although the content often appears gibberish-like upon closer inspection, contrasting with the more consistent CREMMA results. This may be due to the cBAD dataset's greater diversity in script types and its broader historical scope.

The non-Latin datasets (Table 2) show the model's capability in rendering complex manuscript layouts with varying text orientations and marginal notes, as well as the general decor and style of manuscripts, such as Arabic ones. The generated texts, while mimicking authentic letter forms and patterns, remain nonsensical. Notably, in the BiblIA dataset, even without annotations for marginal notes, the model attempts to fill these areas, although not as textual content.

The results from newspaper layouts (Table 2) also show a commendable level of credibility, reproducing complex elements such as columns, titles, and article separators effectively. However, due to the small size of input images, the text often does not materialize as coherent letters-a limitation potentially addressable with higher resolution images.

In summary, while the generated images from regions and baseline masks lack textual content (or, at least, textual content that is semantically valid, and goes beyond filler text), their structural fidelity to medieval handwriting styles is impressive, albeit distinguishable from genuine manuscripts upon human evaluation. This study also explores the impact of the absence of textual content on layout analysis effectiveness. At this step, we do not keep the approach using Text Region masks, less accurate than the one using Baseline masks.

For images derived from characters masks, the outcomes are highly convincing, particularly for Armenian scripts, where distinguishing between authentic and synthetic images is challenging, except for the initials, especially decorated ones, and intonational signs-which are not recognized by the HTR and thus absent from the masks-and decorative elements. But results imitate perfectly Armenian manuscripts, no matter the script considered such as *erkat'agir* (Capital script), *bolorgir* (bicameral non-cursive script) and *notrgir* (cusrive script). In Latin scripts, the model struggles with accurately generating the diverse and often underrepresented abbreviative signs in the datasets, but the result is readable and very accurate.

5.2 About the FID Metric

To evaluate the quality of images generated by Pix2Pix, we use the Frechet Inception Distance (FID) [15], calculated from features extracted by an Inception V3 model trained on ImageNet. This metric assesses the noise level in generated images compared to real reference images. Notably, when real image sets are compared among themselves using FID, scores such as 18.75 for Mix BL, 10.43 for HYE char, and 19.33 for LAT char are observed, suggesting these as new benchmarks for generated images due to the limitations inherent in the Inception model's training.

Throughout the training of three models (Fig. 3) the FID scores remained high, even post-training. Incorporating the aforementioned real image FID scores as reference values, LAT Char and Mix BL align closely with these targets, while HYE Char achieves a lower FID of 10. This discrepancy may stem from challenges in generating illuminations, where the absence of a proper mask sometimes leads to mere attempts at drawing, despite the generated text's high quality.

Qualitatively, by iteration 1000, the text for HYE Char and LAT Char becomes highly legible, and by iteration 1500, it becomes difficult for a non-expert to distinguish these from real documents. Mix BL, achieving a cleaner appearance by iterations 1500 and 2000, remains identifiable as GAN-generated due to minimal constraints in the generation process and the lack of detailed textual information. However, this does not significantly impact the tasks related to layout analysis (see Subsect. 5.3).

These findings highlight FID's limitations when applied to GAN-generated handwritten documents. While effective for evaluating global noise and suitable for manuscripts that focus on line generation, FID does not capture finer textual details. Therefore, enhancing FID assessments with readability criteria-either through a Character Error Rate (CER) threshold, if ground truth and a robust HTR model are available, or by defining thresholds for quality assessments [7]-is crucial to accurately evaluate text generation nuances. Parallel training of a recognizer is generally carried out [11,36,39]. This limitation necessitates the development of a new metric that integrates both visual quality and textual readability, providing a more comprehensive assessment of the generated outputs. To this end, we introduce a hybrid quality score (Q), defined as:

$$Q = \alpha \cdot (1 - \mathrm{Norm}(F)) + \beta \cdot \mathrm{Norm}(R) \tag{1}$$

with:

$$\mathrm{Norm}(F) = \frac{F - F_{min}}{F_{max} - F_{min}} \in [0, 1], \text{ and } \mathrm{Norm}(R) = \frac{R - R_{min}}{R_{max} - R_{min}} \in [0, 1] \tag{2}$$

where F is the FID score, R represents the readability score based on the thresholds good (CER $\in [0, 10]$), acceptable (CER $\in [10, 25]$), bad (CER $\in [25, 50]$) and very bad (CER $\in [50, 100]$) [7] converted into a numerical scale, $\mathrm{Norm}(F)$ and $\mathrm{Norm}(R)$ are the normalized values of FID and readability, respectively, ranging from 0 to 1. The parameters α and β are weights that signify the relative importance of each component-visual quality and readability.

In Eq. 1, $(1 - \text{Norm}(F))$ inversely scales the FID score to align it with the direct proportionality of higher scores indicating better performance, which is consistent with the readability score. This approach allows the comprehensive evaluation of GAN outputs by quantifying both the aesthetics of the manuscripts and the clarity and legibility of their textual content, addressing a crucial gap in existing evaluation methodologies for text-containing images generated by GANs.

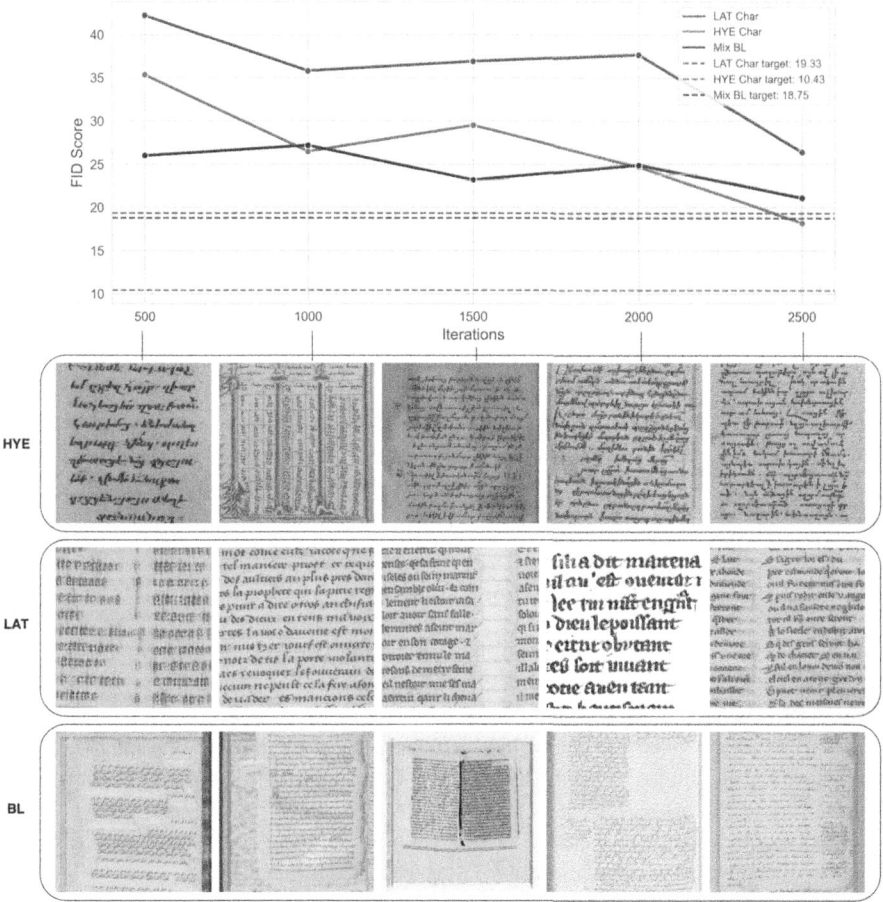

Fig. 3. Evolution of the FID value and corresponding artificial images, generated by the model, for each iteration, for three different datasets, Armenian characters (HYE char), Latin characters (LAT char), and a mixed baseline dataset (Mix BL, including data from CBAD, CREMMA and BADAM)

Applied to HYE Char, the hybrid quality score yields the results reported in Table 3.

Table 3. Hybrid Quality Score Outcomes for HYE Char. Here, $F_{min} = 0$ represents the lowest value of FID indicating a real image, $F_{max} = 200$ is the threshold for non-credible GAN outputs, $R_{max} = 100$, and $R_{min} = 25$. Both α and β are set to 0.5, giving equal importance to visual quality and readability.

Iteration	FID	Readability	Q Score
500	26.01	Poor	0.435
1000	26.49	Acceptable	0.600
1500	29.52	Good	0.926
2000	24.69	Good	0.938
2500	18.21	Excellent	0.955

We observe significant progress between iterations 500 and 1000, and further improvement from iterations 1000 to 1500, even though the FID scores for iterations 500 and 1000 are comparable. Notably, iteration 1500 is penalized by a higher FID of 29.52 (as opposed to 26.49 and 26.01) due to the generation of a stained background-likely the reproduction of a water stain in the manuscript. This effect, while entirely credible and relevant, is treated as noise in FID calculations, despite the presence of damaged images in the real image set. The iterations 1500 and 2000 exhibit very different FIDs but equivalent readability, which minimally impacts the Hybrid score. Although the definition and parameter weighting of this score may require refinement, the metric as defined here underscores the relevance a specific metric tailored to the evaluation of GAN-generated written or handwritten documents.

5.3 Benefits on Layout Analysis

To validate our method, we compared baseline detection results on classic datasets, specifically BADAM and cBAD Simple Track [9], using a CRNN initially developed for BADAM [19]. Data augmentation was applied exclusively in the training set, where we dynamically generated 500 new images per training epoch, replacing the previous batch with in-domain constructed masks based on the dataset.

The results are summarized in Table 4, showing the Precision (P) and Recall (R) percentages for various models of the cBAD competitions.

These results illustrate a significant improvement in performance across both the diverse dataset of cBAD Simple Track and the more complex BADAM dataset. While the augmented CRNN model does not yet surpass the top state-of-the-art models on the cBAD Simple Track, the training pipeline of a CRNN combined with Pix2Pix is notably simpler and computationally less demanding, offering greater flexibility. A key takeaway from this experiment is that despite the lack of textual information (random shapes that imitate manuscript) and the presence of margin noise (blurring effect) in the baseline GAN-generated images, these do not impede the CRNN's performance when applied to real images. A

Table 4. Baseline Detection Results on cBAD Simple Track (ICDAR 2017) and BADAM

Model	Precision (%)	Recall (%)
cBAD Simple Track (ICDAR 2017)		
dhSegment [26]	94.3	93.9
ARU-Net [14]	**97.7**	98.0
Vision U-net	95.1	95.3
CRNN real	94.4	96.6
CRNN fake	69.3	82.8
CRNN + augment 500	97.4	**98.6**
BADAM		
Vision U-net	91.32	85.75
CRNN real	94.1	90.1
CRNN fake	68.1	88.3
CRNN + augment 500	**96.2**	**91.9**

CRNN model trained solely on fake images naturally faces more challenges, yet the recall results are commendable, achieving 82.8% on cBAD and 88.3% on BADAM-outperforming the 2021 U-net version [37] and nearly matching the standard CRNN.

The heterogeneity of annotations in classic datasets-such as differences in annotation strategies by paragraph versus column of text, and variability in the completeness of annotations-complicates direct comparisons at the text region level. For instance, some datasets like BiblIA may not include marginal regions in annotations (see Table 1). We conducted experiments on medieval CREMMA, RASAM, and NewsEye (newspapers), employing both CRNN and YOLO v8 models, with and without generated data, applying a dynamic data augmentation strategy (Table 5).

Despite the application of advanced models and augmentation techniques, the results reveal limited, if any, gains in region detection for both CRNN and YOLO. Notably, there is a significant performance drop in complex datasets like NewsEye, characterized by a high density of regions. The GAN-generated regions often exhibit blurriness and imprecision, and the tendency for the network to generate non-textual information at the exact edge of the mask gives a very artificial appearance to the output (see Subsect. 5.1). Additionally, the GAN fails to assist the CRNN in distinguishing closely situated regions of the same type, often leading to their amalgamation into a single detected region.

One of the interesting use cases of this approach lies less in the augmentation of data for clean and already well-covered documents than in the creation of qualitative data for a very under-resourced target document (due to the script, the support, etc.). A prime example is the manuscript *Torino, Biblioteca Nazionale Universitaria*, L.II.14, which was severely burned in the 1904 library fire. The

Table 5. Region Detection Results on CREMMA Medieval, RASAM, and NewsEye

Model	Precision (%)	Recall (%)
CREMMA Medieval		
CRNN real	91.4	89.6
CRNN fake	32.3	37.9
CRNN + augment 500	91.8	94.6
YOLO v8 real	97.2	96.6
YOLO v8 fake	69.3	74.8
YOLO v8 + augment 500	**97.4**	**97.7**
RASAM		
CRNN real	93.4	92.6
CRNN fake	43.2	21.8
CRNN + augment 500	93.8	93.2
YOLO v8 real	**98.1**	97.9
YOLO v8 fake	71.1	77.4
YOLO v8 + augment 500	98.0	**98.3**
NewsEye		
CRNN real	67.4	58.2
CRNN fake	21.1	19.4
CRNN + augment 500	56.2	54.2
YOLO v8 real	**91.7**	**78.2**
YOLO v8 fake	75.8	73.4
YOLO v8 + augment 500	81.4	76.8

challenge of extracting content from such a degraded document highlights difficulties in layout analysis. To address this, we experimented with generating damaged layouts to enrich the training dataset, which was qualitatively evaluated at the beginning of this paper. Figure 4 presents an original image from the manuscript, and an artificial generation imitating this manuscript using a semantic mask of baselines.

Figure 5 shows that models trained with this augmented dataset, where generated data replaced up to 25% of the real images per epoch, exhibited a significant improvement in recall for detection tasks. Notably, there was a marked decrease in the number of lines fragmented into small pieces compared to models trained solely on real data.

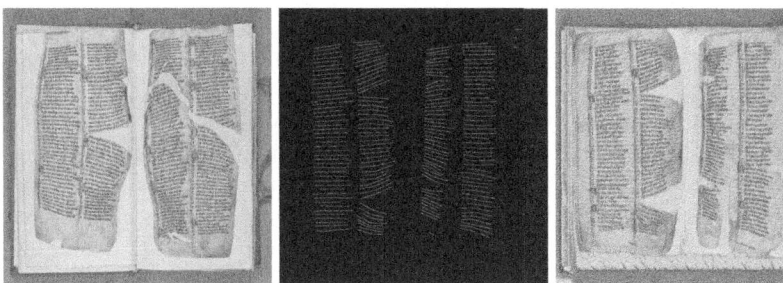

Fig. 4. Generation of damaged document. From Left to Right: Original document (real image), a generated mask, and an artificial image

Fig. 5. Results in layout analysis on a burnt manuscript (Torino, BNU, L.II.14), using only real data in training (on the left) and with data-augmentation using GAN (on the right). Red and green lines indicate indifferently the baselines recognised by the layout analysis models; the model trained without synthetic data tends to over-segment single lines in multiple segments, while the model trained with synthetic data is better at identifying curved or damaged single lines (Color figure online)

6 Conclusion and Further Research

Our current results show the ability of our approach to generate artificial pages with layouts that emulate historical documents such as ancient manuscripts in a credible way for an expert, and to significantly improve layout analysis tasks for documents, especially underrepresented documents with rare layouts, noisy or otherwise damaged. Surprisingly, the approach based on input masks of baselines outperforms approaches based on semantic regions masks.

It is surprising that this architecture, focused on lines and layouts, is actually able, in the absence of a language model or an input text, to generate on occasions realistic looking text, despite its nonsensical nature. Another good surprise is the ability of a fully line-based approach, without the encoding of any semantic zones, to still emulate different types of areas in the layouts (e.g., text, decoration, illumination, titles, etc.).

It could be argued, in that case, than the line-based approach is sufficient, and that the area-based approach is unnecessary, especially given the fact that many datasets do not include relevant information. On the other hand, it is easier and faster for a human to create a handful of fake masks of areas, for on-demand page generations, than a complete set of lines.

Further improvements can be done on three dimensions: data, architecture and evaluation, and in different directions depending on the focus (lines or areas; realistic looking output for humans, or simply usable output for data augmentation tasks). Regarding data, improvements could be obtained by using datasets (or improving existing ones) to include semantic areas whose delimitation more closely matches eventual irregularities, overlaps, etc., of the actual images. Regarding the architecture, and its inputs, we could try to improve the line-based results by giving semantic types to lines based on the type of the zone in which they are contained (for the datasets that have this kind of information).

Finally, regarding the evaluation, due to the limits of native metric for assessing GAN, we propose to proceed in two complementary directions: machine-based and human-based. Machine-based evaluation will be perform by using generated pages as data augmentation for layout-oriented tasks, such as layout recognition, the rationale being that, if the inclusion of synthetic data leads to substantial score improvements, it can be inferred that relevant (and quantifiable) information is contained in the generation. This could be particularly relevant in the case of more complex layouts and small datasets.

Secondly, the qualitative assessment could be performed using a standardised protocol, e.g., historical and fake pages are presented to expert users through an application, in a controlled setting, and the user is asked to assess the nature of the page. The more the assessment of a single page deviates from the average score of a user in assessing the nature of pages (i.e., the more or less the user is "fooled"), the better or worse its score.

These dual evaluations will likely be complementary, as the information used by the machine or examined by expert users might not be the same. For instance, for layout-oriented automated tasks, it is not sure that the precise nature of the text or the quality of the decoration will weigh as much as they do for the expert. For this reason, further improvements could be performed as to maximise one or the other of these scores, or in both directions. The choice between one direction or the other could be determined by the final goal of the generation, i.e. concentrating on data-augmentation, or on the generation of historically plausible pages for human experts, for instance in the perspective of virtual restoration of damaged manuscripts (such as has been recently performed on a text-level for Greek inscriptions [2]).

To conclude, our method demonstrates remarkable versatility in generating convincingly authentic historical data through a streamlined protocol. We noted a general improvement in layout analysis tasks, particularly for resources that are typically underrepresented, thus broadening the scope for studies on less-explored HTR sources. Additionally, the simple yet effective creation of character masks paves the way for generating new texts from previously generated texts, potentially using a language model like an LLM. More broadly, the generated images, whether they depict real text (character mask) or simulated text (line mask), raise intriguing paleographic questions about the composition and perception of features relative to different scripts.

Acknowledgments. This research was funded by the Paris Artificial Intelligence Research Institute (PRAIRIE), under the Human and Social Sciences call, project "Artificial Pasts: Lost Texts and Manuscripts that never were" (P.I. J.B. Camps).

Disclosure of Interests. The authors have no competing interests to declare that are relevant to the content of this article.

References

1. Arroyo, D.M., Postels, J., Tombari, F.: Variational transformer networks for layout generation. In: Proceedings of the IEEE/CVF Conference on Computer Vision and Pattern Recognition, pp. 13642–13652 (2021)
2. Assael, Y., et al.: Restoring and attributing ancient texts using deep neural networks. Nature **603**(7900), 280–283 (2022)
3. Barrere, K., Soullard, Y., Lemaitre, A., Coüasnon, B.: Training transformer architectures on few annotated data: an application to historical handwritten text recognition. Int. J. Doc. Anal. Recogn. (IJDAR), pp. 1–14 (2024)
4. Binmakhashen, G.M., Mahmoud, S.A.: Document layout analysis: a comprehensive survey. ACM Comput. Surv. **52**(6), 109:1–109:36 (2019). https://doi.org/10.1145/3355610
5. Biswas, S., Banerjee, A., Lladós, J., Pal, U.: DocSegTr: an instance-level end-to-end document image segmentation transformer. arXiv preprint arXiv:2201.11438 (2022)
6. Biswas, S., Riba, P., Lladós, J., Pal, U.: DocSynth: a layout guided approach for controllable document image synthesis. In: Lladós, J., Lopresti, D., Uchida, S. (eds.) ICDAR 2021. LNCS, vol. 12823, pp. 555–568. Springer, Cham (2021). https://doi.org/10.1007/978-3-030-86334-0_36
7. Clérice, T.: Ground-truth free evaluation of HTR on old French and Latin medieval literary manuscripts. In: Computational Humanities Research Conference (CHR) 2022 (2022)
8. Clérice, T.: You actually look twice at it (YALTAi): using an object detection approach instead of region segmentation within the Kraken engine. J. Data Min. Digit. Human. (2023)
9. Diem, M., Kleber, F., Fiel, S., Grüning, T., Gatos, B.: cBAD: ICDAR2017 competition on baseline detection. In: ICDAR 2017 – 14th International Conference on Document Analysis and Recognition, vol. 01, pp. 1355–1360 (2017). https://doi.org/10.1109/ICDAR.2017.222
10. Diem, M., Kleber, F., Sablatnig, R., Gatos, B.: cBAD: ICDAR2019 competition on baseline detection. In: ICDAR 2019 – 15th International Conference on Document Analysis and Recognition, pp. 1494–1498 (2019). https://doi.org/10.1109/ICDAR.2019.00240
11. Fogel, S., Averbuch-Elor, H., Cohen, S., Mazor, S., Litman, R.: ScrabbleGAN: semi-supervised varying length handwritten text generation. In: Proceedings of the IEEE/CVF Conference on Computer Vision and Pattern Recognition, pp. 4324–4333 (2020)

12. Gabay, S., Camps, J.B., Pinche, A., Jahan, C.: SegmOnto: common vocabulary and practices for analysing the layout of manuscripts (and more). In: 1st International Workshop on Computational Paleography (IWCP@ ICDAR 2021) (2021)
13. Grüning, T., Labahn, R., Diem, M., Kleber, F., Fiel, S.: Read-bad: a new dataset and evaluation scheme for baseline detection in archival documents. In: 2018 13th IAPR International Workshop on Document Analysis Systems (DAS), pp. 351–356. IEEE (2018)
14. Grüning, T., Leifert, G., Strauß, T., Michael, J., Labahn, R.: A two-stage method for text line detection in historical documents. Int. J. Doc. Anal. Recogn. (IJDAR) **22**(3), 285–302 (2019)
15. Heusel, M., Ramsauer, H., Unterthiner, T., Nessler, B., Hochreiter, S.: GANs trained by a two time-scale update rule converge to a local NASH equilibrium. In: Advances in Neural Information Processing Systems, vol. 30 (2017)
16. Hoyez, H., Schockaert, C., Rambach, J., Mirbach, B., Stricker, D.: Unsupervised image-to-image translation: a review. Sensors **22**(21) (2022). https://doi.org/10.3390/s22218540, https://www.mdpi.com/1424-8220/22/21/8540
17. Isola, P., Zhu, J.Y., Zhou, T., Efros, A.A.: Image-to-image translation with conditional adversarial networks. In: 2017 IEEE Conference on Computer Vision and Pattern Recognition (CVPR) (2017)
18. Kahle, P., Colutto, S., Hackl, G., Mühlberger, G.: Transkribus - a service platform for transcription, recognition and retrieval of historical documents. In: ICDAR 2017 – 14th International Conference on Document Analysis and Recognition, vol. 4, pp. 19–24. IEEE (2017)
19. Kiessling, B., Ezra, D.S.B., Miller, M.T.: BADAM: a public dataset for baseline detection in Arabic-script manuscripts. In: Proceedings of the 5th International Workshop on Historical Document Imaging and Processing, pp. 13–18 (2019)
20. Kiessling, B., Tissot, R., Stokes, P., Ezra, D.S.B.: eScriptorium: an open source platform for historical document analysis. In: ICDAR 2019 – 15th International Conference on Document Analysis and Recognition, Workshops (ICDARW), vol. 2, pp. 19–19. IEEE (2019)
21. Madi, B., Alaasam, R., Shammas, R., El-Sana, J.: Scheme for palimpsests reconstruction using synthesized dataset. Int. J. Doc. Anal. Recogn. (IJDAR) **26**(3), 211–222 (2023)
22. Monnier, T., Aubry, M.: docExtractor: an off-the-shelf historical document element extraction. In: 2020 17th International Conference on Frontiers in Handwriting Recognition (ICFHR), pp. 91–96. IEEE (2020)
23. Muehlberger, G., Hackl, G.: NewsEye/READ OCR training dataset from French Newspapers (18th, 19th, early 20th C.) (2020). https://doi.org/10.5281/zenodo.4293602
24. Najem-Meyer, S., Romanello, M.: Page layout analysis of text-heavy historical documents: a comparison of textual and visual approaches. In: Proceedings of the Computational Humanities Research Conference 2022 Antwerp, Belgium, 12–14 December 2022, pp. 36–54 (2022)
25. Nikolaidou, K., Seuret, M., Mokayed, H., Liwicki, M.: A survey of historical document image datasets. Int. J. Doc. Anal. Recogn. (IJDAR) **25**(4), 305–338 (2022)
26. Oliveira, S.A., Seguin, B., Kaplan, F.: dhSegment: a generic deep-learning approach for document segmentation. In: 2018 16th International Conference on Frontiers in Handwriting Recognition (ICFHR), pp. 7–12. IEEE (2018)
27. Pang, Y., Lin, J., Qin, T., Chen, Z.: Image-to-image translation: methods and applications. IEEE Trans. Multimedia **24**, 3859–3881 (2021)

28. Pfitzmann, B., Auer, C., Dolfi, M., Nassar, A.S., Staar, P.: DocLayNet: a large human-annotated dataset for document-layout segmentation. In: Proceedings of the 28th ACM SIGKDD Conference on Knowledge Discovery and Data Mining, pp. 3743–3751 (2022)
29. Pinche, A.: Cremma Medieval (2022). https://github.com/HTR-United/cremma-medieval
30. Pisaneschi, L., Gemelli, A., Marinai, S.: Automatic generation of scientific papers for data augmentation in document layout analysis. Pattern Recogn. Lett. **167**, 38–44 (2023)
31. Poddar, A., Dey, S., Jawanpuria, P., Mukhopadhyay, J., Kumar Biswas, P.: TBM-GAN: synthetic document generation with degraded background. In: Fink, G.A., Jain, R., Kise, K., Zanibbi, R. (eds.) ICDAR 2023. LNCS, vol. 14188, pp. 366–383. Springer, Cham (2023). https://doi.org/10.1007/978-3-031-41679-8_21
32. Quirós, L.: Multi-task handwritten document layout analysis. arXiv preprint arXiv:1806.08852 (2018)
33. de Sousa Neto, A.F., Bezerra, B.L.D., de Moura, G.C.D., Toselli, A.H.: Data augmentation for offline handwritten text recognition: a systematic literature review. SN Comput. Sci. **5**(2), 258 (2024)
34. Stoekl Ben Ezra, D., Brown-DeVost, B., Jablonski, P., Lapin, H., Kiessling, B., Lolli, E.: BiblIA - a general model for medieval hebrew manuscripts and an open annotated dataset. In: The 6th International Workshop on Historical Document Imaging and Processing. HIP '21, pp. 61–66. Association for Computing Machinery, New York, NY, USA (2021). https://doi.org/10.1145/3476887.3476896
35. Tanveer, N., Ul-Hasan, A., Shafait, F.: Diffusion models for document image generation. In: Fink, G.A., Jain, R., Kise, K., Zanibbi, R. (eds.) ICDAR 2023. LNCS, vol. 14189, pp. 438–453. Springer, Cham (2023). https://doi.org/10.1007/978-3-031-41682-8_27
36. Vidal-Gorène, C., Camps, J.B., Clérice, T.: Synthetic lines from historical manuscripts: an experiment using GAN and style transfer. In: Foresti, G.L., Fusiello, A., Hancock, E. (eds.) ICIAP 2023. LNCS, vol. 14366, pp. 477–488. Springer, Cham (2023). https://doi.org/10.1007/978-3-031-51026-7_40
37. Vidal-Gorène, C., Dupin, B., Decours-Perez, A., Riccioli, T.: A modular and automated annotation platform for handwritings: evaluation on under-resourced languages. In: Lladós, J., Lopresti, D., Uchida, S. (eds.) ICDAR 2021. LNCS, vol. 12823, pp. 507–522. Springer, Cham (2021). https://doi.org/10.1007/978-3-030-86334-0_33
38. Vidal-Gorène, C., Lucas, N., Salah, C., Decours-Perez, A., Dupin, B.: RASAM – a dataset for the recognition and analysis of scripts in Arabic Maghrebi. In: Barney Smith, E.H., Pal, U. (eds.) ICDAR 2021. LNCS, vol. 12916, pp. 265–281. Springer, Cham (2021). https://doi.org/10.1007/978-3-030-86198-8_19
39. Vögtlin, L., Drazyk, M., Pondenkandath, V., Alberti, M., Ingold, R.: Generating synthetic handwritten historical documents with OCR constrained GANs. In: Lladós, J., Lopresti, D., Uchida, S. (eds.) ICDAR 2021. LNCS, vol. 12823, pp. 610–625. Springer, Cham (2021). https://doi.org/10.1007/978-3-030-86334-0_40
40. Wang, H., Wang, Y., Wei, H.: Affganwriting: a handwriting image generation method based on multi-feature fusion. In: Fink, G.A., Jain, R., Kise, K., Zanibbi, R. (eds.) ICDAR 2023. LNCS, vol. 14190, pp. 302–312. Springer, Cham (2023). https://doi.org/10.1007/978-3-031-41685-9_19

41. Xu, Y., Li, M., Cui, L., Huang, S., Wei, F., Zhou, M.: LayoutLM: pre-training of text and layout for document image understanding. In: Proceedings of the 26th ACM SIGKDD International Conference on Knowledge Discovery and Data Mining, pp. 1192–1200 (2020)
42. Zhong, X., Tang, J., Yepes, A.J.: Publaynet: largest dataset ever for document layout analysis. In: 2019 International Conference on Document Analysis and Recognition (ICDAR), pp. 1015–1022. IEEE (2019)

VINALDO

VINALDO 2024 Preface

Document understanding is an essential task in various application areas such as data invoice extraction, subject review, medical prescription analysis, etc., and holds significant commercial potential. Several approaches are proposed in the literature, but datasets' availability and data privacy challenge them. Considering the problem of information extraction from documents, different aspects must be taken into account, such as (1) document classification, (2) text localization, (3) OCR (Optical Character Recognition), (4) table extraction, and (5) key information detection.

In this context, machine vision and, more precisely, deep learning models for image processing are attractive methods. In fact, several models for document analysis have been developed for text box detection, text extraction, table extraction, etc. Different kinds of deep learning approaches, such as GNN, are used to tackle these tasks. On the other hand, the extracted text from documents can be represented using different embeddings based on recent NLP approaches such as Transformers. Also, understanding spatial relationships is critical for text document extraction results for some applications such as invoice analysis. Thus, the aim is to capture the structural connections between keywords (invoice number, date, amounts) and the main value (the desired information). An effective approach requires a combination of visual (spatial) and textual information.

After the success of VINALDO 2023, in the second edition of the VINALDO workshop, we encouraged the description of novel problems or applications for document analysis in the area of information retrieval that has emerged in recent years. On the other hand, we wanted to highlight a particular topic, namely "Multi-view and Multimodal approaches". In fact, the VINALDO workshop aims to combine visual and textual information for document analysis; in this context, multi-view and multimodal methods really have an important advantage in dealing with different types of data. Thus, we encouraged works that combine machine vision and NLP through Multiview or/and multimodal approaches. Finally, we also encouraged works that combine NLP and computer vision methods and develop new document datasets for novel applications. The VINALDO workshop aims to provide a venue for experts from industry, science, and academia to exchange ideas and discuss ongoing research in Computer Vision and NLP for scanned document analysis.

The workshop was organized by Rim Hantach (Engie, France) and Rafika Boutalbi (Aix-Marseille University, France). We had one invited speaker: Juyang Weng, head of the Brain-Mind Institute and GENISAMA, USA. The Program Committee was selected to reflect the interdisciplinary nature of the field. For this second edition, we welcomed two kinds of contributions: short and long papers. We received a total of 11 submissions. Each paper was reviewed by two members of the program committee via EasyChair. A double-blind review was used for the short paper submissions, and the authors were welcome to anonymize their submissions. 5 submissions were accepted and are published here.

August 2024

Rim Hantach
Rafika Boutalbi

Organization

General Chairs

Rim Hantach Engie, France
Rafika Boutalbi Aix-Marseille University, France

Program Committee

Karima Boutalbi LISTIC, Savoie Mont Blanc University, France
Stéphane Ayache LIS, Aix-Marseille University, France
Abilasha Sukumaran LIS, Aix-Marseille University, France
Rafika Boutalbi LIS, Aix-Marseille University, France
Felipe Torres École Centrale Marseille, France

A Multimodal Framework For Structuring Legal Documents

Thibaud Real[✉] and Pauline Chavallard

Doctrine, Paris, France
thibaud.rds@gmail.com

Abstract. Document structuring plays a crucial role in various natural language processing (NLP) tasks, such as information retrieval, and document understanding. It also helps readers to effectively navigate into a structured document with a large amount of textual data. In the legal domain, document structuring is particularly important for creating inter- and intra-document links. In this paper, we present a practical implementation of a multimodal workflow to structure legal documents across various formats. We create a format-agnostic representation of each document (PDF and HTML), that includes layout and textual information. We introduce a multimodal and sequential algorithm to detect titles in each document, and then establish hierarchical relationships among paragraphs using a deterministic algorithm. Our contribution extends to the publication of an open-source dataset, facilitating further exploration in this domain of study, which has received comparatively less attention.

1 Introduction

Many NLP tasks in the legal domain require clean and well-structured documents with precise delimitations between sections. Documents can refer one another, and, in particular, can mention one specific article of one document. A good separation and hierarchy between each paragraph is needed in order to perform links, extract relevant information, and perform other tasks (Fig. 1).

Traditionally, manual effort was, and is still required to define sections (articles, chapters, ...) and to establish hierarchical relationships among these. However, with the emergence of machine learning, researchers have explored automated approaches to tackle these tasks. But in the legal domain, where even a single mistake can be costly, these approaches are not yet widely adopted.

Document structuring is a complex task that poses several challenges such as handling domain-specific documents, identifying implicit titles, adapting to the diversity of documents and formats, and resolving ambiguous paragraph relationships. Even for a human, it can be hard to distinguish a title from a content, especially for non legal specialists. Our research specifically concentrates on legal documents, which introduces additional challenges. One such difficulty

Fig. 1. Random examples from our dataset.

arises from the common practice of quoting other documents within legal content, leading to the inclusion of structural elements from those documents within the initial document, thereby creating an ambiguous structure.

Although some efforts have been made to explore document structuring, existing approaches have been relatively limited. Previous research can be divided into multiple areas. Some approaches have been made towards the layout analysis of documents, where deep learning models have been developed: LayoutParser [1], LayoutLM models [2–4], and LayoutXLM [5], a multilingual version that could be applied to french documents. They can be used to extract various elements such as lists, tables, images, titles, and subtitles from documents. However these models need to be applied to very specific documents independently, not generalizing well to new type of documents. Also, these approaches are not well suited to process HTML documents. Furthermore, initial experiments on the DocLayNet dataset seem to indicate that these models are not optimal in predicting titles [6].

A second group of approaches try to tackle the problem by extracting Table of Contents (ToC) from documents to parse the hierarchical structure of sections and subsections (for example [7]). But many legal documents do not have an explicit table of contents.

Other research have aimed to determine the logical structure of documents at a fine-grained level [8]. While this approach have shown promise, it presents a level of granularity (and thus complexity) that is not needed in our work.

[9] proposes a method to generate a table of contents even if it does not exist in the document, but this method deals with the title detection task on each paragraph locally and separately and only focuses on PDF documents.

Our research seeks to develop a generic framework that can effectively structure different types of documents in various formats (HTML, PDF image or text), using both layout and textual information, and leveraging the sequential information present in the document to improve the title identification part.

2 Problem Description

The goal of this study is to identify all titles within a document and establish the hierarchical relationships among all paragraphs.
To accomplish this, we employ the following steps:

Division Into Rext Units. We initially extract and divide the document into smaller text units, typically one text unit is one line of text). Each line is given a unified representation, ensuring consistent metadata information regardless of the input format.

Title Identification. Next, we identify and extract the text units recognized as titles. This step involves a binary classification process where each line is categorized as either a title or not. It's worth noting that although rare, there may be cases where a line contains both a title and the beginning of its associated content, but we did not encounter such examples in our dataset.

Hierarchical Structure Creation. Finally, we construct a hierarchical structure of these titles by assigning a level l_i to each title i. The convention used is as follows: if $l_i = k$, it means that title i is at a higher level compared to $l_i = k+1$. Thus, if title i is at level k and title $i+1$ is at level $k+1$, it implies that title $i+1$ is nested within title i. Assessing the level of each title's content becomes straightforward, as each content is placed at the same hierarchical level as its preceding title.

3 Dataset

We are releasing our dataset, named *DoctrineAI/legal_document_structuring*[1] on the HuggingFace platform. Further information about its usage and format can be found on the corresponding dataset card. Our intention is that the accessibility of this dataset fosters collaboration and innovation in the comprehension and utilization of visually structured documents.

3.1 Dataset Construction

For the purpose of our research task, we generated a customized dataset comprising French legal documents. The reason for creating our own dataset stems from the absence of existing French datasets for the described task. Our dataset encompasses a diverse range of documents, characterized by varying formats and layouts. Examples include international tax agreements and various preparatory documents, all of which are accessible through open data sources. Further details on how we achieved a unified representation for each document are provided in the next section.

---

[1] https://huggingface.co/datasets/DoctrineAI/legal_document_structuring,.

3.2 Dataset Information

Our dataset includes a diverse collection of documents written in French, which are available in both PDF and HTML formats. Table 1 provides detailed information about the dataset.

To label and detect titles of each document, we employed regular expressions. However, the task of defining what constitutes a title posed certain challenges. In certain cases, distinguishing whether a particular text line should be classified as a title or not proved to be difficult. In such cases, we decided to use the titles that would typically be included in a table of contents. For instance, we excluded headers from the table of contents, as they are not considered titles.

It is important to note that despite our efforts with the regular expressions, some inaccuracies still persist in our dataset.

To ensure a rigorous evaluation process, we carefully selected a test set consisting of 39 documents of diverse types. These documents were manually annotated to avoid any potential issues during the evaluation.

Table 1. Dataset Information

Format	Source	#Text units	%Titles	#documents
PDF	Tax Agreements	67101	15.8	190
	Notice	12691	23.7	100
	Impact Study	164765	11.9	143
HTML	Debate	19780	2.6	50
	Impact Study	30812	24.8	84
	Meeting	141	2.8	5
	Motives	1825	15.5	50
	Notice	2099	12.1	5
	Report	63495	18.5	100

4 Constructing a Format Agnostic Representation

The first step of our framework consists in creating a format-agnostic representation of each document.

To do so, we tried to gather layout information that is accessible in any input format for each line of text, describing the visual representation of each line of text. All extracted information is given in Table 2. Features are normalized so that values do not depend on the tool and format used.

We cut each PDF document into lines using pdftotext from poppler tool and gather layout information (bold, italic, position) for each line.

For HTML documents, we cut each document into the more granular units of texts possible, using typical tags that represent unit of texts : *div, p, h1, h2, h3...* and get the same metadata information as for PDF documents. We did not use the tag names as extra features because we wanted to have identical information for both PDF and HTML documents. For each unit of text, we extracted layout information (bold, italic, position) using selenium tool.

Layout information have been normalized in order to avoid having differences when using different tools. For example, the left indent is relative to the length of the page.

Table 2. Extracted attributes for each text unit.

Text unit attribute	Description
is_bold	Is the text bold?
is_italic	Is the text italic?
max_height	Line maximum height
left_indent	Left margin

5 Detecting Titles

5.1 Baseline

Features. Initially, we attempted to implement a simple algorithm that incorporates ideas from various papers. We developed an algorithm that, for each text unit, first extracts textual and visual features, as well as some information of its preceding and following text units.

All constructed features for the baseline are detailed in Table 3.

Table 3. Features of the baseline title detection model.

	Feature name	Description
Layout	is_bold	Is the text bold?
	is_italic	Is the text italic?
	max_height	Max line size of the text
	left_indent	Length from the page to the left part of the text
	next_text_same_style	Does next text unit have same style (bold, italic)?
	prev_text_same_style	Does previous text unit have same style (bold, italic)?
	most_freq_left_indent	Does the text unit have the most frequent indent?
Textual	startswith_upperchar	Does the text start with an upper character?
	endswith_punctuation	Does the text ends with (.,;!?)?
	enswith_semicolon	Does the text ends with ":"?
	startswith_special	Does the text ends with a special character (-, ., ...)?
	text_length	Number of characters in the text
	upperchar_percent	Percent of upperchar character in the text.
	contains_title_name	Does the text contain a section name (*Chapitre...*)?
	startswith_num_upper	Does it starts with a number then an upper char. ?
	is_understandable	Is the text understandable (excludes url, etc.)?

Classification. The baseline consists of a simple classifier that uses previous features to determine whether each text unit is a title. The model used is a Random Forest [10] with a hundred trees. It classifies each text unit separately. The only sequential information it can access concerns style features. It evaluates whether the following and previous text units have the same style as the current text unit. We did not perform hyperparameter optimization.

5.2 Proposed Model

To enhance the previous model and achieve a more generalized model, we turned to an implementation based on recurrent neural networks (RNNs).

A primary consideration was to create a model that wouldn't demand excessive resources. This is particularly relevant due to the potential length of legal documents, which can extend to over 500 pages in some cases. It is therefore intersting to consider models with lower resource usage. Consequently, we opted to design an LSTM-based [11] architecture instead of more computationally demanding language models like BERT or recent decoder-only architectures.

Our model uses both text and visual features, and it also exploits the sequential nature of the document in order to predict which of the input paragraph of the documcnt is a title.

The model architecture is represented on Fig. 2. It consists of the concatenation of a character-CNN [12] representation and a pre-trained word embedding derived from a Word2Vec [13] model trained on French legal documents (note that while this specific embedding may not be publicly available, similar results can be achieved using open-source word embeddings). It is then followed by a BiLSTM network and an attention layer [14]. This allows obtaining a representation of each paragraph. Each paragraph is then passed into a final BiLSTM network followed by a linear classification layer, predicting for each paragraph whether it is a title or not. In the final BiLSTM layer, we incorporate a concatenated representation of both paragraph features and handcrafted features.

Our intuition behind the use of a character-CNN embedding is for example to capture the paragraph numbering information frequently utilized in titles. The use of pre-trained word embeddings helps in identifying conjugated verbs (often indicating non-title text) or specific words commonly present or absent in titles. Additionally, the attention layer enables the model to focus on particular words and enhances its interpretability.

The proposed model is trained for a maximum of 15 epochs with an early stopping condition to prevent overfitting. We used standard parameters and did not try to perform hyper-parameter tuning.

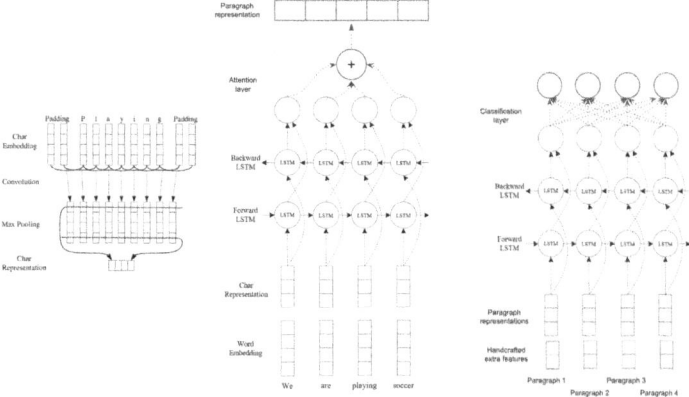

Fig. 2. Title detection model architecture.

6 Construction of Hierarchical Relationships

In order to establish the hierarchical relationships between titles, we introduce a global deterministic algorithm that is less influenced by local errors encountered during the title detection step. However, it is worth noting that the construction of the hierarchy was not the primary focus of this study. Additional comprehensive and in-depth evaluations may be necessary to further design and evaluate this part of the study.

Refer to the Fig. 3 below for an illustration of the described process.

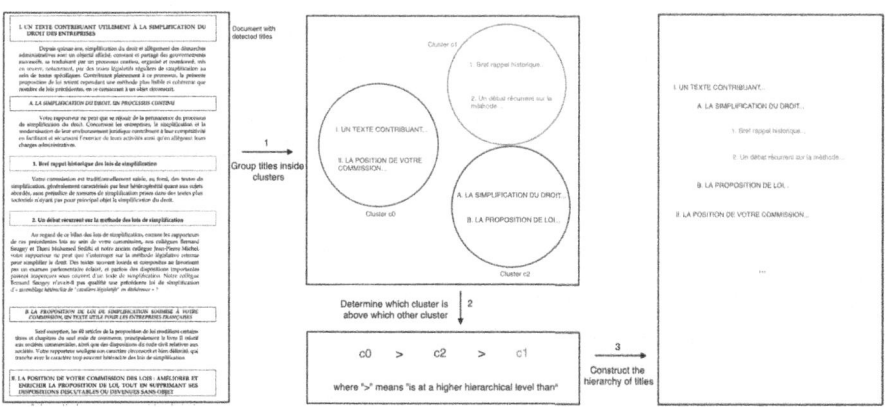

Fig. 3. Illustration of the steps of the hierarchy algorithm. After step 3, the text indentation implicitly denotes the inclusion of one section within another.

6.1 Titles Clustering

The process begins with grouping titles with similar properties into clusters. This step is based on similar naming, resulting in grouping titles such as "Chapter X" ($X \in N$) together, or similar numbering, resulting in grouping titles such as "I", "II", "III" and "IV" together. The underlying concept is that these clusters have consistent behavior within a given document. Specifically, one cluster will always be positioned above another cluster within the document structure.

The features used to group titles together are: *is_italic*, *is_bold* and *normalized_numbering*. For the first two features, please refer to Table 2. The *normalized_numbering* feature is constructed by first extracting the numbering of the title (if the title does not have a numbering, the feature value is empty). The numbering is then normalized by converting each number to its first corresponding value. For example:

- The normalized numbering of "5" is "1" (since "1" is the first number in a basic numbering).
- The normalized numbering of "III" is "I" (Roman numeral "I" is the first in a Roman numeral numbering).
- The normalized numbering of "B.3.c" is "A.1.a" (alphabetical numbering starts with "A" and sub-sections use lowercase letters).

Finally, titles with identical values for all three features are grouped together to form a cluster.

6.2 Relationships Between Clusters

Once the titles have been grouped into clusters, the next step involves determining the existing relationships among these clusters. To evaluate these relationships, a first naive option is to examine the order of appearance of each cluster. If cluster A precedes cluster B in the reading order, it indicates that cluster A is positioned hierarchically above cluster B. However, an incorrect detection of a title may lead to an erroneous hierarchy. To resolve this issue, a more global analysis of the clusters of the document is conducted. This analysis is performed by considering the average occurrence of titles from cluster B between two consecutive titles from cluster A. If the titles from cluster B appear more often, on average, between two titles from cluster A than the opposite, then it indicates that cluster A should be placed hierarchically above cluster B. While this approach may not be theoretically accurate in certain specific cases, empirical observations on our dataset demonstrate its effectiveness.

6.3 Constructing The Final Hierarchy

The final step involves iterating through each title and utilizing the established relationships to construct the complete hierarchy of titles.

Calculating the level for titles located within or lower in the hierarchy than the previous title is straightforward: for each title i, if title $i-1$ is higher than title i, then $l_i = l_{i-1} + 1$.

However, if title i is positioned (hierarchically) above title $i-1$, determining the hierarchical level becomes ambiguous, requiring an examination of all preceding titles until a title k is encountered that belongs to the same cluster as title i. In such cases, the hierarchical level of title i is set equal to the hierarchical level of title k: $l_i = l_k$.

7 Evaluation

7.1 Title Detection

Baseline. The initial model described earlier achieved an overall F1 score of 0.893 for the title detection task. However, after a qualitative study, it is interesting to note that the model still makes significant errors on certain types of documents (see detailed results on Fig. 4) and does not generalize well to documents from sources not included in the training dataset. For instance, when the model was applied to a document from the European Court of Human Rights (a document type not present in the training data), numerous errors were observed.

Our Model. The proposed model performs well for the title detection task for which we obtain a F1 score of 0.953 on our manually labelled test dataset. The precise evaluation for each type of document is detailed on Fig. 4. We observe that the labels quality (involved due to rule-based labelling), the number of documents in the dataset, and the diversity of titles seem to be the main factors explaining the differences between the results of different types of documents. On tax agreements documents, for example, we observe that titles are quite always the same (low diversity of titles) and very well labelled using regular expressions and thus the model performs very well on this type of document.

A qualitative evaluation also highlights that this model generalizes interestingly well to other (and unseen) new type of documents that are not present in training data. We tested the model on new types of documents (e.g. European legislation documents) and the results seemed quite promising. Further evaluation on unseen data would be necessary to proceed with this part (Table 4).

Table 4. F1-score on manually labelled test dataset for the title detection task.

Format	Source	Baseline	Our Model
PDF	Tax Agreements	0.982	0.99
	Notice	0.816	0.952
	Impact Study	0.723	0.959
HTML	Debate	0.609	0.839
	Impact Study	0.773	0.976
	Meeting	1	1
	Motives	0.954	0.969
	Notice	0.844	0.809
	Report	0.895	0.93
	All	0.893	**0.953**

7.2 Title Hierarchy

In order to evaluate the constructed hierarchy of each document, we conducted a comparative analysis of the hierarchical levels associated with each title across all documents. The score was computed by comparing the (flattened) lists and calculating the mean squared error. This metric was chosen based on the underlying notion that greater penalties should be imposed for disparate levels. Our model obtains a score of 0.792 when evaluating only the hierarchy part (based on ground-truth titles), and a score of 1.044 when evaluating title detection followed by the construction of the hierarchy. Our experiment also revealed that the designed algorithm performs better than the naive option consisting in examining only the order of appearance of each cluster to determine the relationships.

Please note that evaluation of the hierarchy was not the primary focus of this study, and additional analysis may be needed for this part.

7.3 Interpretability

The model's attention layer allows to have a way to understand the model's predictions, adding an interpretability layer to the title detection process. It allows to identify which words the model focuses on when predicting titles. A concrete example is given on the Fig. 4. We observed that the model concentrates on conjugated verbs to predict that a paragraph is not a title. It focuses on the word "article" to predict that all articles are section titles. It should be noted that our architecture does not allow us to easily visualize the importance of layout features (bold, italics, centered, etc.).

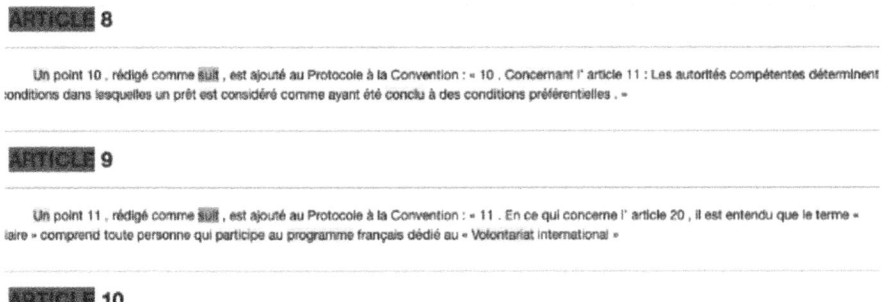

Fig. 4. A concrete example demonstrating the model interpretability. The model intensifies its focus on words highlighted in red. Specifically, we can see that it does focus onto usual words included in lots of titles (like "article") and conjugated verbs (like "suit"). Interestingly, it does not seem to focus on numbers.

8 Deployment Considerations

Our framework has been successfully deployed into production on several type of documents. The title detection model can run on CPU or GPU. In order to make it faster, we process batches of paragraphs (of one same document) simultaneously. The processing time of the hierarchical algorithm is negligible compared with the processing time of the title detection model. The processing time highly depends on the size of the document. For very long documents, such as reports from the national assembly containing hundreds of pages, the framework typically processes only a few documents per second. However, for very short documents, it can process up to hundreds of documents per second.

The framework we propose is divided into three distinct stages. Although our title detection model is not universally applicable to all types of documents due to its lack of accuracy for certain new document types, we easily substituted it with manual rules when necessary. The remaining steps of the framework remain unchanged and do not require any modifications.

9 Conclusion

In this paper, we have introduced a format-agnostic framework for constructing the hierarchical structure of a legal document by leveraging both textual and layout information. Our approach begins by extracting text and similar layout information from HTML and PDF documents. The title detection step involves an efficient recurrent deep learning model that uses textual and visual features while considering the sequential nature of the document. Our model achieves great results on our manually labelled dataset consisting of diverse french legal documents. Moreover, our model is (partially) interpretable by its nature, bringing some useful debugging information. Additionally, we provide explanations of a deterministic algorithm for establishing hierarchical relationships among titles within the document. Further research includes trying other recent models such as BERT-based sequential classification model or decoder-only archiecture models, despite the inherent challenges posed by the length of certain legal documents.

References

1. Shen, Z., et al.: LayoutParser: a unified toolkit for deep learning based document image analysis. arXiv preprint arXiv:2103.15348 (2021)
2. Xu, Y., Li, M., Cui, L., Huang, S., Wei, F., Zhou, M.: LayoutLM: pre-training of text and layout for document image understanding. In: Proceedings of the 26th ACM SIGKDD International Conference on Knowledge Discovery & Data Mining (2020). https://doi.org/10.1145/3394486.3403172.
3. Xu, Y., et al.: LayoutLMv2: multi-modal pre-training for visually-rich document understanding. arXiv preprint arXiv:2012.14740 (2022)
4. Huang, Y., Lv, T., Cui, L., Lu, Y., Wei, F.: LayoutLMv3: pre-training for document AI with unified text and image masking. arXiv preprint arXiv:2204.08387 (2022)

5. Xu, Y., et al.: LayoutXLM: multimodal pre-training for multilingual visually-rich document understanding. arXiv preprint arXiv:2104.08836 (2021)
6. Guillou, P.: Document understanding model (finetuned LayoutXLM base at paragraph level on DocLayNet base). https://huggingface.co/pierreguillou/layout-xlm-base-finetuned-with-DocLayNet-base-at-paragraphlevel-ml512
7. Nguyen, T.T.H., Doucet, A., Coustaty, M.: Enhancing table of contents extraction by system aggregation. In: 14th IAPR International Conference on Document Analysis and Recognition (ICDAR), pp. 242–247 (2017). https://doi.org/10.1109/ICDAR.2017.48.
8. Koreeda, Y., Manning, C.D.: Capturing logical structure of visually structured documents with multimodal transition parser. arXiv preprint arXiv:2105.00150 (2021)
9. Bentabet, N.I., Juge, R., Ferradans, S.: Table-of-contents generation on contemporary documents. arXiv preprint arXiv:1911.08836 (2019)
10. Breiman, L.: Random Forests. Mach. Learn. **45**, 5–32 (2001)
11. Hochreiter, S., Schmidhuber, J.: Long short-term memory. Neural Comput. **9**(8), 1735–1780 (1997)
12. Zhang, X., Zhao, J., LeCun, Y.: Character-level convolutional networks for text classification. arXiv:1509.01626 (2015)
13. Mikolov, T., Chen, K., Corrado, G., Dean, J.: Efficient estimation of word representations in vector space. arXiv:1301.3781 (2013)
14. Bahdanau, D., Cho, K., Bengio, Y.: Neural machine translation by jointly learning to align and translate. arXiv:1409.0473 (2014)

Reformulating Key-Information Extraction as Next Sentence Prediction for Hierarchical Data

Ashish Kubade, Prathyusha Akundi, and Bilal Arif Syed Mohd

Perfios Software Solutions Private Limited, Bengaluru, India
ashish.kubade@gmail.com

Abstract. We present a reformulation of the Key-Information Extraction (**KIE**) problem from document images, as a Next-Sentence Prediction (**NSP**) task for identifying information in hierarchically structured data. KIE implemented as a Key-Value extraction task, is limited to one-to-one (single key mapping to single value) information extraction and thus does not apply to hierarchical information e.g. information present in complex semi-structured or unstructured tables. The Visual-Question-Answering (**VQA**) approach tries to solve information extraction from such semi-structured formats, but use visual information extraction backbone architectures along with heavy language models. In the proposed work, we use only a backbone language feature extractor for semantic entity extraction. Unlike, the four entity types in FUNSD ('question', 'answer', 'header' and 'other'), for semi-structured tabular information we define additional classes that define hierarchical elements, like column-header, table-footer, cells, merged-cell, table-summary etc. For these additional entities, we define hierarchical relations like a tuple of entities {table-header entity, column-header entity, row-header entity} that point to the unique entity referred as a value-entity. We treat tuple-entity and value-entity as two sentences and formulate the task of finding how likely is the value-entity to follow the tuple-entity. Empirically, we show that the proposed method, called as, Tuple-Value Identification (**TVI**), can exhaustively identify all the information in the hierarchical structures. Additionally, TVI also opens up for the potential use for Table Structure Recognition (TSR) for scanned documents in bank statements or medical bills, where the narration columns span multi-lines and is challenging for existing TSR systems.

Keywords: Key Information Extraction · Next Sentence Prediction · Hierarchical Information Extraction · Semantic Entity Recognition · LayoutLM

1 Introduction

Identification of Key-Value pairs or tables representing structured data within documents plays a pivotal role in information extraction and document understanding tasks. However, extracting key-value pairs from document images remains a challenging problem due to diverse contents, variations in layout and templates, handwritten texts, noise introduced during scanning, etc. Using generic architectures like LayoutLM that leverages the spatial information (2D spatial coordinates of the text of the tokens) from the document, MTL-FoUn [10] solves the KIE problem in three stages viz, word grouping, entity labelling and entity linking. Instead of using 2D layout information directly, BROS [2] proposed the use of relative positions of text blocks with area-masking strategy. GeolayoutLM [7] uses vision models to extract geometric information between tokens for learning the geometric layout representation explicitly. Though these architectures achieve excellent performance for Relation Extraction (**RE**) task, the relations that these architectures define are one-to-one, meaningly, single key-entity is expected to link to single value-entity. So, a standard RE solution can not be applied to data formats having hierarchical nature. A Question-Answering (QA) based key-value pair extraction approach has been proposed in KVPFormer [3], where they use transformer based encoder-decoder model for key-value pair extraction.

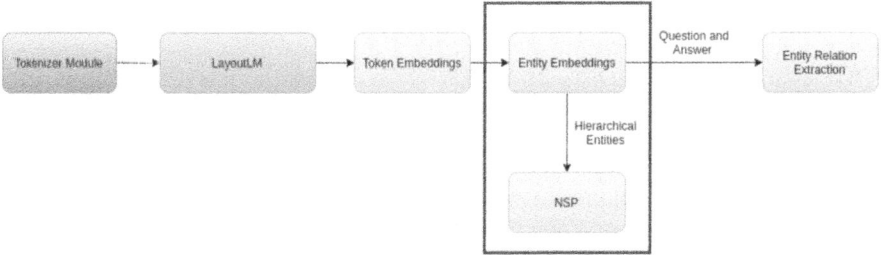

Fig. 1. An overview of KIE solution using TVI Architecture. The TVI module is highlighted with red box. (Color figure online)

Tables on the other hand, can contain complex information in hierarchical fashion and are difficult to extract without using text and visual features. For example, architectures like TableFormer [8], TSRFormer [6] work great on tabular or hierarchical tables. However, their performance degrades rapidly over poor quality scanned documents as these architectures are based on visual information which is vulnerable to changes in presence of noise. With above limitations in extracting information as purely key-value identification approach or purely table-structure identification approach, we present Tuple-Value Identification (TVI) that identifies the information present as key-value pairs along with hierarchical information present in the form of semi-structured tables.

We reformulate the task of key-value extraction in the presence of hierarchical structure as next-sentence prediction (NSP) task. As shown in Fig. 1, similar to MTL-FoUn [10], our approach starts with LayoutLM generating token embeddings followed by a grouping head. The token groups, referred as entities, are passed through an entity classifier model. However, unlike MTL-FoUn [10], we use 15 classes for entity classification, in addition to {other, header, question, answer} that are available in FUNSD [5]. We introduce labels that are representative of hierarchy in the information like table-header, column-header, cells, merged-cell, table-summary etc. As shown in Fig. 1, our TVI architecture has standard entity relation extractor similar to MTL-FoUn [10] that identifies links between simple question and answer entities, whereas the NSP head takes in the hierarchical entity elements and extracts information that can be presented with the help of directed keys or tuples.

To handle the extended set of entity labels, we filter in elements like column-header, row-header etc. and build sets of 4 entities that form our tuples. We treat first column-header entity as 'question-header' and entity elements under first column as 'question-elements'. Subsequent column-header elements act as 'answer-headers'. For each combination of {question-header, question-element, answer-header} called as 'tuple-entity', that uniquely identify a piece of information, will be the 'answer-element' or 'value-entity'. The tuples and their respective values are passed to Next Sentence Prediction module that predicts the likelihood of value-entity following the tuple-entity.

Since, our method (TVI) identifies standard one-to-one key-values pairs (as in KIE) along with hierarchically organized information that generally occurs in semi-structured tables, our novel approach of identifying both these kinds of information from the documents makes TVI generic towards any document type. Since, currently there are no publicly available standard datasets that provides annotation in the format TVI expects, we use AWS Table service [1] to get the table elements and build the hierarchical annotations from SROIE [4]. However, we do provide our results on FUNSD [5] and CORD v2 [9] datasets which provide annotations for standard RE task. We evaluate the performance of TVI against state-of-the-art methods like LayoutLM [11], BROS [2], GeoLayoutLM [7], MTL-FoUn [10] and show the effectiveness of the method on RE task.

Major contributions of this paper are as follows:

– Reformulation of Key Information Extraction for hierarchical data as Next Sentence Prediction Task
– A data pipeline using AWS Table service to generate Tuple-Value annotations for data generation

2 Method

Inspired from MTL-FoUn [10], we solve the problem in three steps. As shown in Fig. 2, we use LayoutLM as our semantic and layout information encoder. The sub-word token embeddings are passed to a word grouping model that generates word groups. These word groups, referred as Entities, are used for Semantic Entity Recognition (**SER**) task. We formulate the SER task as a 15 class classification problem which can be seen as: {question, answer, header, other} as in FUNSD [5], {table-header, table-title, table-summary, table-footer, merged-cell, cell, column-header}, etc. for hierarchical or semi-structured tables, {question-without-answer, answer-without-question} for the entities where the question do not have answers, or answers that are explicitly mentioned and lastly {organization-name, organization-logo-text} for the entities that represent the organization details, etc.

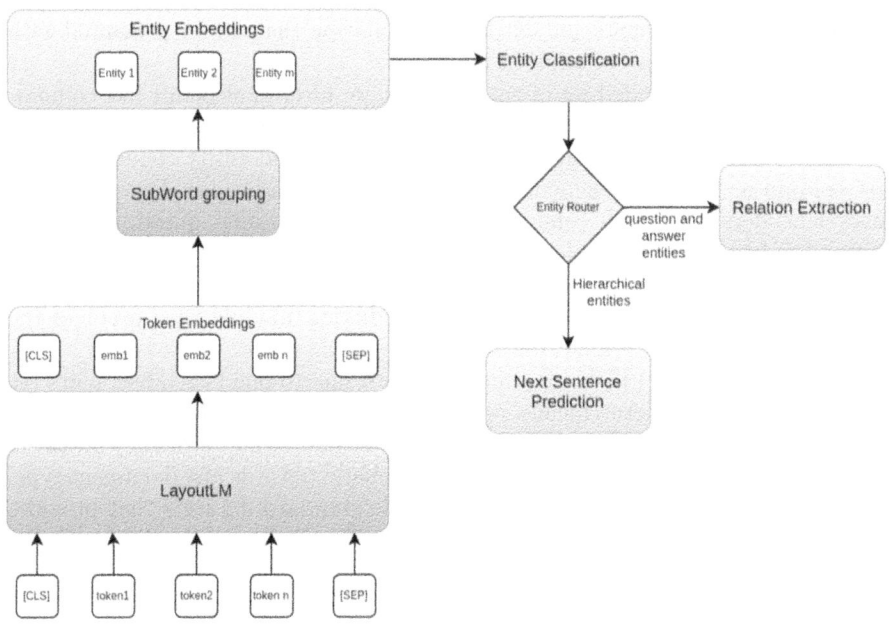

Fig. 2. Overall architecture of Tuple-Value Identification Solution

Since our SER formulation can differentiate between entities, we apply a entity filter and bifurcate the SER task via two paths. The standard question and answer entities are sent to SER model. Entities that form hierarchical information and are required for Tuple-Value Formulation are passed to NSP model.

2.1 Sub-word Grouping

We attach a set of linear layers as a Sub-word to refine the LayoutLM embedding that are used to compute hinge loss during training. During inference, similar to MTL-FoUn [10], we use Agglomerative clustering to predict sub-words that can be grouped together. Since, sub-word grouping is the first task during inference, any errors made during sub-word group formation propagate to SER and ER tasks, particularly on the multi-line key entities. Hence we enhance the one-layer architecture used by MTL-FoUn [10] and use 3 layered perception to get final sub-word token features.

2.2 Semantic Entity Recognition

Semantic Entity Recognition starts with aggregating the LayoutLM embeddings of the sub-words that comprise that entity. We use 3 layered perception with 15 output classes to define the SER head. SER head receives the average sub-word group entities and returns the class probabilities for the 15 classes. As mentioned in Sect. 2, these classes are {question, answer, header, other, table-header, table-title, table-summary, table-footer, merged-cell, cell, column-header, question-without-answer, answer-without-question, organization-name, organization-logo-text}. We use standard cross-entropy loss while training the model.

2.3 Relation Extraction (RE) Task

For the entities that occur in one-to-one, i.e. like questions and answers in FUNSD [5], we again follow the approach by MTL-FoUn [10] where we learn a parameter matrix and find the likelihood of a question-entity mapping to a particular answer-entity.

2.4 Formulation of Next Sentence Prediction

From the output of SER, the entities of the type 'column-header' are used to build tuple-value pairs. One of the 'column-header' entities is considered as 'question-header' while the others as 'answer-headers'.

For example, as shown in Fig. 3, for a multi-column table, we define the first column-header as question-header e.g. Entity 'Earnings'. Subsequent column-headers are treated as answer-headers e.g. 'Full' and 'Actual'. The elements in the first column form as question-entities whereas the elements in subsequent columns are used as answer-entities, e.g. 'BASIC', 'HRA', 'CONVEYANCE', 'SPECIAL ALLOWANCE', etc. and '18754', '6452', etc. respectively. A few of the tuple-value pairs are then formulated as {'Earnings', 'BASIC', 'Full'} -> '18954', {'Earnings', 'MEDICAL ALLOWANCE', 'Actual'} -> '1250'

We use 2 layered perceptron that acts as a binary classifier. We use sequential training for each tuple-value pair and accumulate the cross-entropy loss over all the tuple-value pairs in the document.

In a document, with multiple tables, in order to limit the scope of entities to respective table, we use AWS service to extract tables. Entities that belong to same table are only considered while generating tuple-value pairs. Also, to identify hierarchical information structures, we select only those tables where at least 3 'column-header' entities are present. In this way, we make sure that we have one 'question-header' and at least 2 'answer-headers'. In cases where we have only 1 or no 'answer-headers', the problem degrades to simple one-to-one information extraction.

An example of tables extracted from AWS Table service [1] are shown in Fig. 4 where we see three tables have been extracted. For one of the table, we show the tuple-pair formulation in Fig. 3.

Earnings		Full	Actual
BASIC		18854	18654
HRA		7502	7502
CONVEYANCE		1600	1600
MEDICAL ALLOWANCE		1450	1250
SPECIAL ALLOWANCE		6452	6452
Total Earnings: Rs.		35958	35558

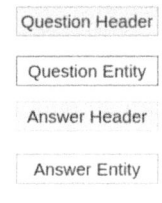

Fig. 3. Tuple-Value pair formulation

Earnings	Full	Actual	Deductions	Actual
BASIC	18754	18754	PF	1800
HRA	7502	7502	PROF TAX	250
CONVEYANCE	1600	1600		
MEDICAL ALLOWANCE	1250	1250		
SPECIAL ALLOWANCE	6452	6452		
Total Earnings: Rs.	35558	35558	Total Deductions: Rs.	2050
Net Pay for the month (Total Earnings - Total Deductions):	33508			
(Rupees Thirty Three Thousand Five Hundred Eight)				

This is a system generated payslip and does not require signature.

Fig. 4. Tables extracted from AWS Table service. Green boxes denote the table boundaries. (Color figure online)

3 Datasets

For the standard SER and RE task, we use FUNSD [5], SROIE [4] and CORD v2 [9] datasets. FUNSD dataset offers annotations for SER and RE task with 4 classes viz {question, answer, header and other} for SER. CORD v2 [9], however, is more complex towards SER as it offers hierarchical annotations as 5 superclasses and 30 sub-classes among them. However, both these datasets do not have annotations for RE task for hierarchical information. Hence to train NSP model, we annotate samples from SROIE [4] (with the help of AWS Table service), which has tabular information that can be formulated in tuple-value pairs.

4 Experiments and Results

We conduct 2 experiments to validate the performance of TVI. First we train TVI independently on FUNSD for both SER and RE tasks. Since our architecture is mostly inspired from MTL-FoUn [10], we can see similar performance over SER (refer Table 1). Secondly, since CORD v2 has 30 labels for SER, we train our SER model for 30 classes. The results are available in Table 1. Alongside, we also mention the model sizes in terms of number of parameters. TVI being a light-weight model compared with GeoLayoutLM [7] and BROS [2], has lesser inference time and is potentially a better choice for real-time applications.

Table 1. Results on FUNSD dataset for the standard SER task

Method	Model Size (Number of params)↓↓	F1↑↑	
		FUNSD	CORD v2
MTL-FoUn	**113M**	85.0	–
BROS	340M	84.53	97.28
GeoLayoutLM	399M	**92.86**	**97.97**
TVI-FUNSD (Ours)	114M	83.2	87.3

For qualitative analysis of SER model, we show a sample result in Fig. 5. We can see that model has identified entities of types 'question' (labelled as 'Que' and highlighted in brown), 'answer' (labelled as 'Ans' and highlighted in red). Boxed with peach color represent entities of class 'column-header' denoted with 'CH', and so on. In the left table, model has predicted three column-headers and hence NSP module will get triggered. However, in the right table, we see we have only two column-headers, hence a simple RE solution will be applied.

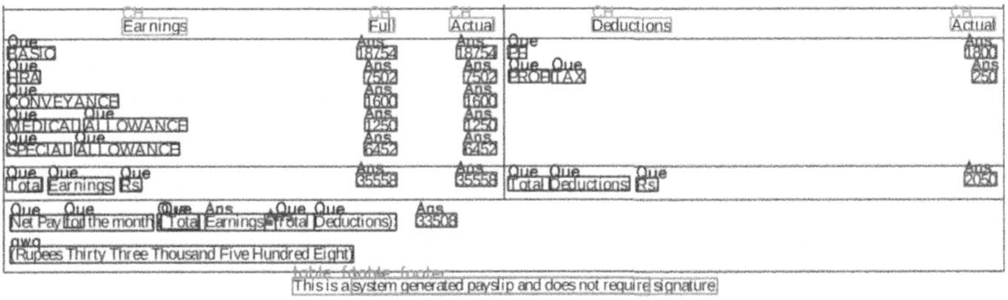

Fig. 5. Sample Visual results with the 15 class SER. Each box represents one of the semantic entities.

For the qualitative results on RE task, we use only question, answer entities without additional context. Figure 6 represents the results on the RE task where entities enclosed in blue boxes represent 'question' entities, and green boxes represent 'answer' entities. We see that for hierarchical information present in left-table, model performance is poor as there are two potential answer entities for each question entity. Whereas the RE task becomes easier for one-to-one (one question to one answer) mapping.

For TVI results on hierarchical data, we represent the visual results on one of the test sample that we have chosen from publicly available salary slip template in Fig. 7.

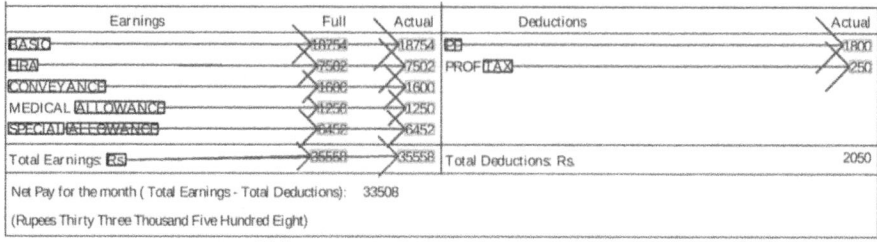

Fig. 6. Visual results on RE task.

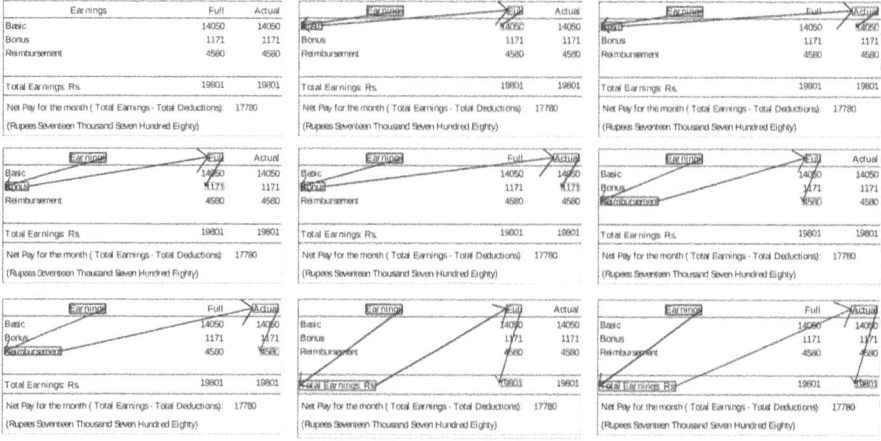

Fig. 7. Sample Visual results with TVI. Top left image in the grid represents the input sample. For the remaining samples, the direction of the arrow represent the order of entities considered in the tuples. NSP correctly predicts the last entity given the context of previous three entities.

5 Conclusion and Future Work

In this paper, we proposed the approach of key-value extraction for the hierarchical information. Our approach involves leveraging existing methods for identifying key-value pairs for simple information contents and Next Sentence prediction where the information can be extracted as directed key-value pairs in the form of tuple-values. With our initial experiments, we show that the proposed architecture TVI can achieve similar performance on SER task with even though our goal was to build a Next Sentence Prediction model. For RE, we believe with extensive research, TVI could extract key-value pairs from any general document, mimicking visual-question answering models with much lesser number of parameters. With TVI, we also seek the possibility of using it for Table Structure Identification where each of the cell entity can be extracted using SER and the cell associativity can be built using NSP module from TVI.

References

1. Aws tables. https://docs.aws.amazon.com/textract/latest/dg/how-it-works-tables.html. Accessed 19 Apr 2024
2. Hong, T., Kim, D., Ji, M., Hwang, W., Nam, D., Park, S.: Bros: a pre-trained language model focusing on text and layout for better key information extraction from documents. In: Proceedings of the AAAI Conference on Artificial Intelligence, vol. 36, pp. 10767–10775 (2022)
3. Hu, K., Wu, Z., Zhong, Z., Lin, W., Sun, L., Huo, Q.: A question-answering approach to key value pair extraction from form-like document images. In: Proceedings of the AAAI Conference on Artificial Intelligence, vol. 37, pp. 12899–12906 (2023)
4. Huang, Z., Chen, K., He, J., Bai, X., Karatzas, D., Lu, S., Jawahar, C.V.: ICDAR2019 competition on scanned receipt OCR and information extraction. In: 2019 International Conference on Document Analysis and Recognition (ICDAR), pp. 1516–1520 (2019). https://doi.org/10.1109/ICDAR.2019.00244
5. Jaume, G., Ekenel, H.K., Thiran, J.P.: FUNSD: a dataset for form understanding in noisy scanned documents. In: 2019 International Conference on Document Analysis and Recognition Workshops (ICDARW), vol. 2, pp. 1–6. IEEE (2019)
6. Lin, W., Sun, Z., Ma, C., Li, M., Wang, J., Sun, L., Huo, Q.: TSRFormer: table structure recognition with transformers. In: Proceedings of the 30th ACM International Conference on Multimedia, pp. 6473–6482 (2022)
7. Luo, C., Cheng, C., Zheng, Q., Yao, C.: GeoLayoutLM: geometric pre-training for visual information extraction. In: Proceedings of the IEEE/CVF Conference on Computer Vision and Pattern Recognition, pp. 7092–7101 (2023)
8. Nassar, A., Livathinos, N., Lysak, M., Staar, P.: TableFormer: table structure understanding with transformers. In: Proceedings of the IEEE/CVF Conference on Computer Vision and Pattern Recognition, pp. 4614–4623 (2022)
9. Park, S., et al.: CORD: a consolidated receipt dataset for post-OCR parsing. In: Workshop on Document Intelligence at NeurIPS 2019 (2019)
10. Prabhu, N., Jain, H., Tripathi, A.: MTL-FoUn: a multi-task learning approach to form understanding. In: Barney Smith, E.H., Pal, U. (eds.) ICDAR 2021. LNCS, vol. 12917, pp. 377–388. Springer, Cham (2021). https://doi.org/10.1007/978-3-030-86159-9_27
11. Xu, Y., Li, M., Cui, L., Huang, S., Wei, F., Zhou, M.: LayoutLM: pre-training of text and layout for document image understanding. In: Proceedings of the 26th ACM SIGKDD International Conference on Knowledge Discovery & Data Mining, pp. 1192–1200 (2020)

HPSegNet: A Method for Handwritten and Printed Text Separation in Document Images

Yu Chao[1,2], Changsong Liu[1,2(✉)], Liangrui Peng[1,2], and Yanwei Wang[1,2]

[1] Department of Electronic Engineering, Tsinghua University, Beijing, China
chaoy21@mails.tsinghua.edu.cn, lcs@tsinghua.edu.cn
[2] Beijing National Research Center for Information Science and Technology (BNRist), Beijing, China

Abstract. The separation of handwritten and printed text in document images is an important task in the optical character recognition (OCR) research field. It is still a challenging problem to separate overlapped handwritten and printed text lines in images of complex documents including examination papers, legal documents, etc. In this paper, handwritten and printed text separation is formulated as a pixel-level document image segmentation task. Firstly, a modified Transformer-based model is designed for pixel-level document image segmentation. Secondly, a residual feature bypass is incorporated into the model to further exploit high-resolution features. Finally, a loss function combining focal loss and dice loss is designed to tackle the problem of imbalanced distributions of different classes. Experimental results on both a public English document image dataset and a self-built Chinese document image dataset have demonstrated the effectiveness of the proposed method.

Keywords: Handwritten Text · Printed Text · Document Image Segmentation · Deep Learning

1 Introduction

Document images, such as historical documents [1], prescription notes [2], legal and court-issued documents [3], business contracts [4], and test papers [5], contain both handwritten text and printed text. When people write text on a printed document, the handwritten text usually overlaps printed pixels. To facilitate information extraction from these document images, it is a fundamental step to separate handwritten text pixels from the whole image. This task is challenging due to the variety of writing styles and complex structures of the documents.

Recently deep learning-based methods [6] have achieved promising performance on image semantic segmentation tasks. Dutly et al. [7] use U-net fully convolutional architecture with conditional random field postprocessing to solve the task of printed and handwritten text separation at the pixel level. Gholamian and Vahdat proposed a new dataset SignaTR6K with pixel-level annotations [8], and used convolution blocks as the fine-feature-path(FFP) to assist U-net segmenting the handwritten and printed pixels.

The advantage of convolutional neural networks(CNN) such as fully convolutional network(FCN) is that high-level features and patterns are captured, but the detailed information is irreversibly lost in the process of down-sampling. In addition, CNN naturally has a limited receptive field and cannot establish long-distance connections of global information, which limits the performance of segmentation. Currently, no research is working from the perspective of expanding the receptive field and making the model focus more on the key regions.

We aim to classify all the pixels into 4 classes: handwritten, printed, overlapped, and background. Segmenting overlapped pixels is the key point in this task. Although this kind of pixel accounts for a tiny part of all pixels, it plays a critical role in the subsequent recognition task. Due to the nature of scattered distribution, these pixels are extremely difficult to classify. Following the work of Xie at al. [9], we provide a solution, *Handwritten and Printed text Segmentation Network (HPSegNet)* to separate printed text and handwritten text in document images, and have achieved better performance compared with the previous method.

The main contributions of this paper are as follows:

a) We designed a novel deep-learning-based method for separating handwritten and printed pixels in complex document images. A network structure including a Transformer-based encoder and a light-weight decoder is proposed to address the semantic features, and a bypass of residual shortcut is added to augment high-resolution features.
b) A loss function combining focal loss and dice loss is designed to tackle the problem of imbalanced distributions of different classes.
c) Experiments on the public English document image dataset SignaTR6K and a self-built Chinese document image dataset verified the effect of the proposed method. Our method has outperformed other reported methods on the SignaTR6K dataset .

2 Related Works

2.1 Traditional Methods

Some traditional methods have explored this problem. These methods usually consist of several steps including image preprocessing, text localization, feature extraction, and classification. Zagoris et al. [10] propose an approach to use the Bag of Visual Words (BoVW) model with Scale-Invariant Feature Transform (SIFT) features extracted from blocks of interest in binarized document images, and classify the boxes of interest as Handwritten text or Printed text or Noise by using support vector machine(SVM). Belaïd et al. [11] also use SVM to classify extracted pseudo words, and use K-NN for pseudo-word grouping. The above traditional methods do the classification task at the box level, however, the handwritten text boxes inevitably overlap the printed boxes, resulting in a serious disruption in the subsequent OCR process. In this paper, we solve this problem by pixel-level segmentation, with the overlapped pixels classified into a distinct category *overlapped pixels*(**OP**).

2.2 Deep Learning Based Methods

Some researchers have studied the segmentation of handwritten text and printed text by deep learning methods, such as [8,12–14]. Some only focus on binary classification, in which a pixel is classified into handwritten pixels or not [13]. Some adopted a 3-classes formulation, with the classes of handwritten, printed and background [12,14]. However, these methods fail in the overlapped area. Gholamian and Vahdat proposed the 4-classes method [8] and Mixed Feature Model(MFM)model. MFM is a deep learning method to segment handwritten and printed pixels, which combines a Fine Feature Path (FFP) with a Semantic Segmentation Path (SSP) and improves performance by capitalizing on both high-level and low-level features. SSP is a U-net style model with 4 convolutional blocks, in which the output feature information is lost during down-sampling, so FFP is introduced to assist the segmentation head with low-level high-resolution features. FFP is also composed of four convolutional blocks, but unlike SSP, FFP maintains the same feature resolution as input by controlling the convolution stride and zero-padding. However, the features extracted solely by 3 * 3 convolutions are not sufficient to represent the difference between handwritten pixels and printed pixels. When the semantic information of the feature itself is rich enough and the resolution is high enough, the effect of the branch is very limited. In this paper, we discussed and verified the impact of three kinds of different bypasses on our model.

3 Methods

We define the handwritten and printed text segmentation as a 4-class pixel classification task. For a given document D, the pixels are classified into *Printed Text (PT), Handwritten Text(HT), Overlapped Pixels(OP)* and *Background(BG)*.

3.1 System Framework

The system framework of HPSegNet is shown in Fig. 1. This structure consists of a Transformer-based encoder, a lightweight decoder, and a residual feature bypass. Given an input image $X \in \mathbb{R}^{H \times W \times C}$, the encoder maps it to a hidden embedding h, and the decoder maps the hidden embedding h to the output image $Y \in \mathbb{R}^{H \times W \times C}$. However, the hidden embedding is not a traditional 1D vector, because semantic information exists at least 3 dimensions in the image. Thus we still use the convolution method to embed the features, and the output embedding has 3 dimensions H_i, W_i, C_i.

We introduced the hierarchical Transformer encoder [9], which can output multi-scale feature embeddings, including high-resolution features and low-resolution features. Taking into account the global attention of the upper level and the local attention of the lower level, the attention layer in the transformer

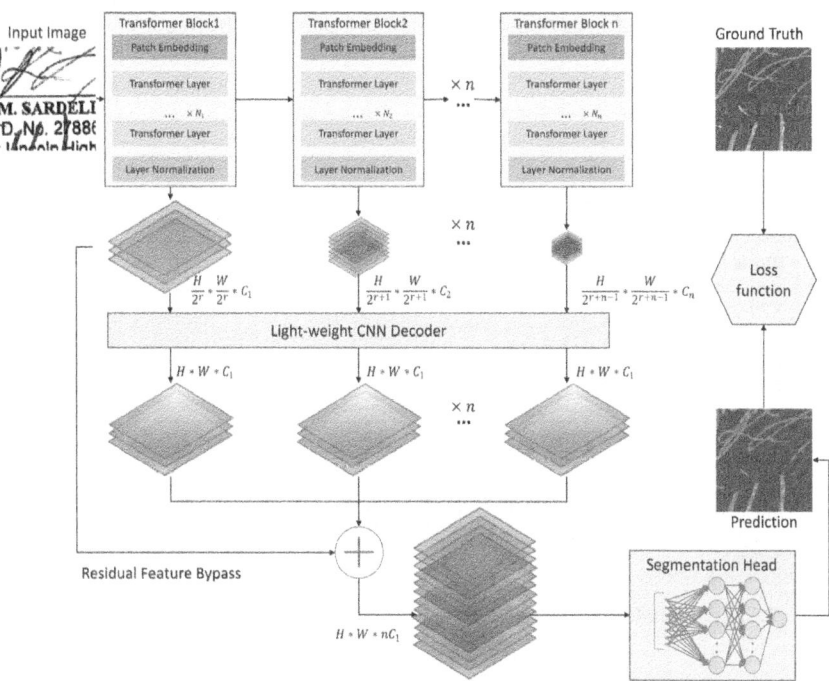

Fig. 1. System Framework. Different from other transformer-based encoders, we deleted the positional embedding module, and our Transformer-based encoder outputs multi-scale features other than a single feature.

block, which follows the CNN-based patch embedding layer, effectively expands the receptive field compared to single CNN layer.

In the decoder part, we combine the 1×1 convolution and bilinear interpolation to merge all layers' output information and process the dense prediction problem, which not only reduces the computational load but also effectively fuses the features of each layer.

Additionally, we add a parallel residual feature bypass, a simple shortcut starting at a transformer block output. The shape of the letters and the thickness of the strokes on the handwritten text are very sensitive, thus when segmenting document images, high-resolution features are necessary to refine the segmentation result. We introduced this part to assist the encoder-decoder model in obtaining more low-level detail information.

Hierarchical Transformer Encoder. The general encoder for text pixel segmentation applies a Mix Transformer Encoder structure(MiT) [9]. The specific structure is shown in Fig. 3. Unlike ViT [15], which can only output a single fixed-resolution feature embedding, MiT can output multi-scale features, including high-resolution features in shallow layers and low-resolution features in deep

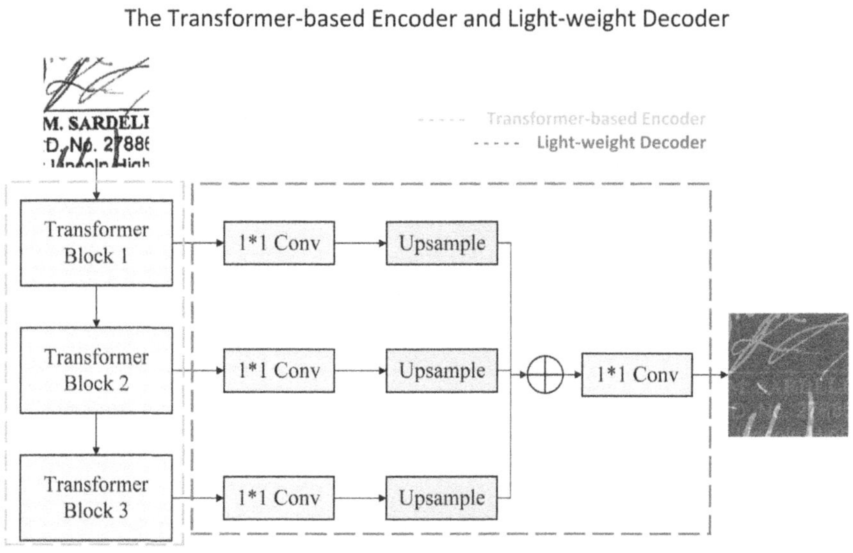

Fig. 2. The Transformer-based Encoder and Light-weight Decoder Framework

layers. During the patch embedding stage, we use a simple convolution method to get the patches and map them to embeddings, then attention is introduced to capture the global information for each patch. Additionally, Mix-FFN(Mix-Feed-Forward-Network) is introduced to take into account the location information brought by zero padding, as is shown in Fig. 3.

Lightweight CNN-Based Decoder. Although the Transformer encoder structure can effectively extract the feature information of the input image, it also brings the problem of high computation cost. We introduce a decoder composed of simple CNN layers to receive the multi-scale features output by the encoder, effectively use the feature information, and output the discrimination results for each pixel. In our experiment, the decoder accordingly uses 1×1 convolution to receive and unify the channel dimension, which is the first CNN layer. The second layer is used to up-sample the features to the resolution of the first transformer block output feature and concatenate them, and the last layer predicts the segmentation mask of the input image. The decoding process of the overall decoder can be summarized as the Fig. 2.

3.2 Design of the Transformer-Based Encoder

The Resolution of the Features. In the baseline model, the encoder outputs a series of feature embeddings with different resolutions. More specifically, when the size of the input image is $H \times W \times 3$, the size of the output feature embedding is $\{\frac{H}{4} \times \frac{W}{4} \times C_1, \frac{H}{8} \times \frac{W}{8} \times C_2, \frac{H}{16} \times \frac{W}{16} \times C_3, \frac{H}{32} \times \frac{W}{32} \times C_4\}$, where $\{C_1, C_2, C_3, C_4\}$

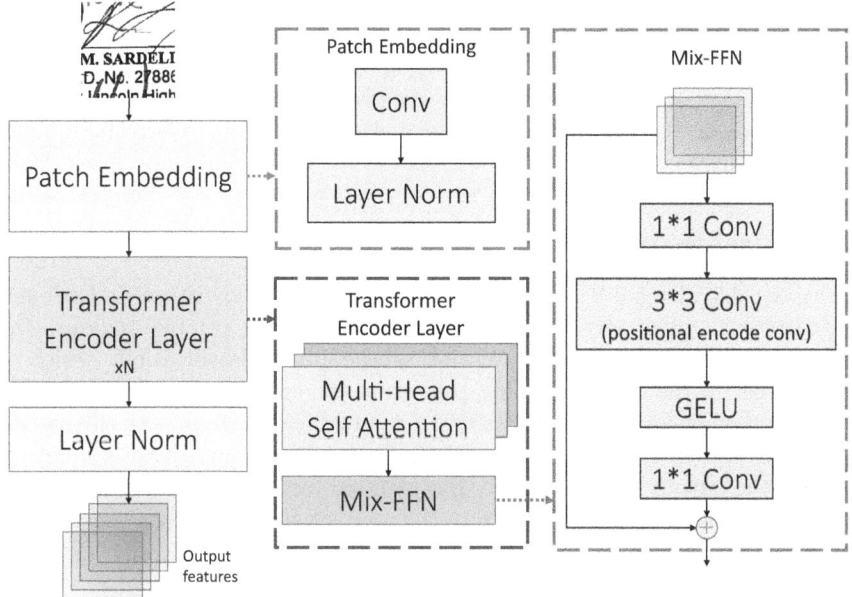

Fig. 3. Transformer Block Structure.

denote the number of embedding channels. Among them, the shallow features are high-resolution features, while the deep features are low-resolution features.

However, text pixel classification is a dense classification problem. The segmentation result and its downstream OCR process are very sensitive to the segmented text strokes. When the input image and features are down-sampled, the detailed information is inevitably and rapidly lost, which is an irreversible process. To solve this problem, we attempted to explore a series of different output feature resolutions by controlling the convolution stride in the first transformer block, as is shown in Table. 2. We do not down-sample the input image to a $\{\frac{H}{4} \times \frac{W}{4} \times C_1\}$ resolution, but maintain the output feature resolution the same as the original image $\{H \times W \times C_1\}$, or down-sample to only $\{\frac{H}{2} \times \frac{W}{2} \times C_1\}$.

The Number of Transformer Blocks. The number of transformer blocks is directly related to the computation cost of the network. Although in theory, the deeper network can lead to better results, the actual situation is not entirely consistent with that. Firstly, when the number of parameters becomes larger, the real-time performance of segmentation tends to be worse, and the hardware requirements for inference become higher, which makes it more difficult to deploy to the downstream OCR processes. In addition, as the network becomes deeper, the accuracy of the model will reach saturation or even degradation. Therefore, adopting an appropriate number of transformer blocks for text segmentation tasks is necessary.

In the baseline model, there are four transformer blocks, with encoder layers of {3,6,40,3}in each block. We experimented with different configurations of transformer blocks and analyzed the output features of each block, which are shown in Table. 2 and Fig. 4. Finally, we removed the fourth transformer block, and the encode layers configuration of our method is {3,6,40}. Correspondingly, the decoder only receives the output features of the three blocks and fuses them.

3.3 Residual Feature Bypass

The encoder-decoder model performs well in distinct handwritten and printed areas. However, when the handwritten strokes overlap printed regions, the detailed low-level features are essential to the pixel classification, which is inevitably lost in the down-sampling process of the encoder-decoder model. To fully utilize the low-level features, we add a residual feature bypass to the model, which starts at the output of the first transformer block, and directly concatenates with the output of the decoder. The residual feature bypass can effectively introduce the high-resolution feature to the segmentation head.

Fig. 4. The internal feature representation in a model with 4 Transformer blocks and a Fine-Feature-Path [8]. The color shows the heatmap values of feature maps. The original image from SignaTR6K dataset is also shown. (1)–(4) present the output features of the 1st–4th transformer blocks respectively in our baseline model. (5)–(8) present the output feature of the 1st-4th convolution blocks respectively in FFP. The output of the 1st Transformer block has more detailed high-resolution features than those in the FFP, which inspires us to incorporate a residual feature bypass instead of the FFP.

Different from fine-feature-path (FFP) [8], which consists of 4 convolution modules with residual connection, the residual feature bypass directly adopts the high-resolution feature of the first transformer block without down-sampling. FFP uses zero padding to ensure that the output feature resolution remains the same as the original image resolution. However, the semantic information of CNN extraction is not enough as the transformer block, which can be seen in Fig. 4. The residual feature bypass demonstrates a good improvement, as discussed later in the paper.

3.4 Loss Function

Document images naturally have the characteristic of extremely uneven pixel distribution, with background pixels accounting for the majority and text pixels accounting for a relatively small portion. In the training dataset of SignaTR6K, the percentage of BG, PT, HT, and OP are $\{79.94\%, 5.04\%, 14.19\%, 0.83\%\}$ respectively. Therefore, when segmenting pixels, text pixels tend to be divided into background pixels. We considered this issue when designing the loss function, and we intend to solve it from two perspectives: firstly, selecting the appropriate loss function, and secondly, assigning appropriate weights to pixel categories, not only to effectively handle the problem but also to stabilize the network while training and inference. The loss function consists of two parts, L_f and L_d.

The first loss function is focal loss [16], which is based on cross-entropy loss, and aims to force it to the harder and incorrectly classified samples by applying a modulating factor $(1-p_t)^\gamma$. p_t indicates the difficulty of classifying a sample, that is, higher p_t represents higher confidence in the predicted result, which means the sample tends to be easier to classify. Therefore, focal loss increases the weight of harder samples in the loss function, which helps to improve the accuracy of the predicted result:

$$L_f = -(1-p_t)^\gamma log(p_t) \qquad (1)$$

where

$$p_t = \begin{cases} \hat{p}, & y=1 \\ 1-\hat{p}, & otherwise \end{cases} \qquad (2)$$

Additionally, we also applied the dice loss [17] to alleviate the long-tail problem in mixed document images. Dice loss is defined as:

$$\begin{aligned} L_d &= 1 - \frac{2|X \cap Y|}{|X|+|Y|} \\ &= 1 - \frac{2TP}{2TP+FP+FN} \\ &= 1 - \frac{2 \times precision \times recall}{precision + recall} \end{aligned} \qquad (3)$$

In this definition, $|X|$ and $|Y|$ refer to the predicted sample number of categories X and Y respectively. In multi-classification problems, it can be extended to the following form:

$$L_d = 1 - \frac{1}{N_{cls}} \sum_{i=1}^{N_{cls}} \frac{2 \times precision_i \times recall_i}{precision_i + recall_i} \quad (4)$$

It can be seen that dice loss is equal to $1 - Fscore$, so it naturally has the optimization goal of maximizing F-score, and focuses more on the mining of the foreground area, while focal loss can solve the problem of unstable training of dice loss, so we combine the two kinds of loss for network training. In this multi-pixel category segmentation task, we also need to consider the category balance. We have counted the distribution of handwritten pixels and printed pixels in the training dataset, and found that the number of text pixels is far smaller than the number of background pixels. If cross-entropy loss is used without considering the extremely unbalanced category, the segmentation effect will be horrible. We assign pixels with a computed weight:

$$w_i = -\frac{N_{cls}}{log n_i \sum_{i=1}^{N_{cls}} log n_i}, i = 1, 2, ..., N_{cls} \quad (5)$$

where N_{cls} is the number of pixel categories, and n_i is the number of pixels of class i. Such weight distribution can effectively solve the problem of hard classification, making the weight value of each category as close as possible to 1, and train the network at the normal level. Finally, the overall loss is calculated as follows:

$$L_{final} = -\alpha \sum_{i=1}^{N_{cls}} gt \cdot w_i \cdot (1-p_t)^\gamma log(p_t)$$
$$+ \beta(1 - \frac{1}{N_{cls}} \sum_{i=1}^{N_{cls}} \frac{2 \times w_i \times precision_i \times recall_i}{precision_i + recall_i}) \quad (6)$$

In our experiment, we assign α and β as 1.0.

4 Experimental Results

4.1 Experiment Setup

Our experiments are carried out on a public English document image dataset and a self-built Chinese document image dataset. The public dataset is used to compare our method with other methods.

The public SignaTR6K [8] dataset is generated from 200 original legal documents in English from Thomson Reuters Legal Content Services [8]. The SignaTR6K dataset has pixel-level annotations, which not only annotate background, handwritten, and printed pixels but also annotate overlapped pixels. Each original document image is cut into samples with size of 256 × 256 pixels, thus there are 6257 images in the SignaTR6K dataset, of which 5169 are used for training, 530 are used for validation, and 558 are used for testing.

We used the mmsegmentation [18] codebase for the experiment and used a server with a single V100 for training, which is configured with 32 GB of memory. We trained the network model on the SignaTR6K dataset for 40k iterations.

In the training process, we use *random flip and photo metric destruction* as the data preprocessing pipeline. During the training process, we set the batch size to 16 for the 1-block and 2-blocks model, and 1 for 3-blocks and 4-blocks model. We used AdamW as the optimizer, and initialized the learning rate to 0.00006, then used the "poly" learning strategy with 1.0 as the power, and conducted 1500 times of warm-up learning rate linear iteration with 0.0000001 as the ratio. Finally, we use the mean Intersection of Union (mIoU) score to judge the segmentation effect of handwritten pixels and printed pixels.

4.2 Experimental Results on the Public English Document Image Dataset

Comparison with Other Methods. We compared our results with other existing methods on SignaTR6K. Table 1 shows the differences between our method and other existing methods. FCN-light and MFM methods are convolutional networks. The expansion of the receptive field by the transformer has led to HPSegNet's performance far exceeding that of CNN-based methods by 7.77%.

The visualization of the test result is shown in Fig. 5.

Table 1. The experiments of comparison with other state-of-the-art methods. *"PT"*, *"HT"*, and *"BG"* refer to *Printed Text, Handwritten Text*, and *Background* respectively.

Model	IoU(PT)	IoU(HT)	IoU(BG)	mIoU
FCN-light [19]	64.55%	89.21%	98.39%	84.05%
SSP-InceptionV3 [8]	72.82%	92.44%	98.73%	87.99%
MFM(FFP+SSP)-InceptionV3 [8]	73.10%	92.66%	98.77%	88.18%
Baseline(4 blocks)	81.79%	87.27%	97.51 %	88.86%
HPSegNet(3 blocks+Residual)	**92.7%**	**98.06%**	**99.8%**	**96.87%**

Selection of the Number of Transformer Blocks in the Model. The baseline model of HPSegNet adopts 4 transformer blocks with the number of transformer encoder layers as {3,6,40,3}. We first analyze the output feature of every block. Fig 4 shows the feature heatmaps. The first feature map contains more fine-grained information, and the second feature map focuses more on the boundary of both printed and handwritten texts. The third feature map focuses more on the printed region, but it is difficult to explain the significance of the region that the fourth feature map focuses on. Thus we assume that at the third transformer block, the network's ability to understand image semantic information has reached its upper limit. We designed a series of experiments with the number of blocks as {1,2,3,4}. The results in Table. 2 show that the 3-blocks configuration can achieve the best performance.

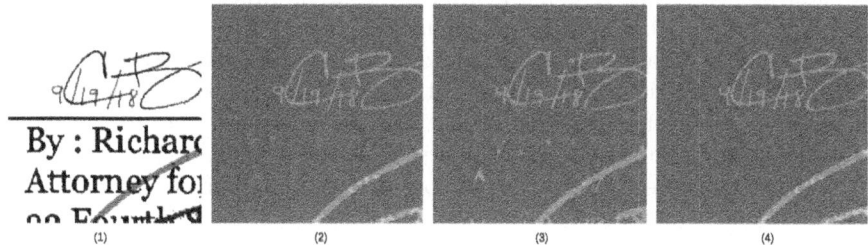

Fig. 5. The comparison of baseline method and HPSegNet(3Block-1Resolution-1ResidualFeatureBypass). (1) the input image; (2) the ground truth; (3) the baseline method result; (4) our method result. The blue pixels represent background pixels. The red pixels denote printed pixels. The green pixels are handwritten pixels. The yellow pixels denote overlapped pixels. (Color figure online)

Influence of the Feature Resolution. In baseline model, the output feature resolutions of four transformer blocks are defined as $\{\frac{H}{4} \times \frac{W}{4}, \frac{H}{8} \times \frac{W}{8}, \frac{H}{16} \times \frac{W}{16}, \frac{H}{32} \times \frac{W}{32}\}$. We investigate the different choices of the output feature resolution of the first transformer block. The results in Table. 2 shows the improved performance by setting the output feature resolution of the first transformer block as $\{H \times W\}$.

Effect of the Bypass. The feature extracted result in Fig. 4 shows the limitation of FFP convolution bypass. The transformer block effectively expands the receptive field and focuses more on key regions. Therefore, we attempt to directly introduce residual feature bypass from the output of the transformer block or the input image. Results in Table 3 show that the direct residual feature bypass outperforms the FFP convolution bypass. In addition, when the feature resolution of the first transformer block is high enough, the role of the bypass is relatively limited.

Table 2. Performance with different configurations of the Transformer-based encoder. "Resolution" refers to the output feature resolution of the first transformer block. The baseline model has 4 Transformer blocks, and the "Resolution" is $\{\frac{H}{4} \times \frac{W}{4}\}$.

Block Num	Params	Resolution	IoU(PT)	IoU(HT)	IoU(BG)	mIoU
1	0.293M	$\{H \times W\}$	83.44%	94.05 %	99.8%	92.43%
		$\{\frac{H}{2} \times \frac{W}{2}\}$	87.13%	95.05 %	99.63%	93.94%
		$\{\frac{H}{4} \times \frac{W}{4}\}$	74.83%	83.09 %	96.89%	84.94%
2	4.007M	$\{H \times W\}$	88.56%	96.52%	99.8%	94.96%
		$\{\frac{H}{2} \times \frac{W}{2}\}$	90.31%	96.59%	99.67%	95.52%
		$\{\frac{H}{4} \times \frac{W}{4}\}$	82.6%	87.59 %	97.55%	89.25%
3	52.286M	$\{H \times W\}$	**92.59%**	**98.00%**	**99.82%**	**96.80%**
		$\{\frac{H}{2} \times \frac{W}{2}\}$	91.29%	97.04%	99.7%	96.01%
		$\{\frac{H}{4} \times \frac{W}{4}\}$	82.43%	87.63 %	97.59%	89.22%
4	81.971M	$\{H \times W\}$	92.03%	97.92%	99.82%	96.59%
		$\{\frac{H}{2} \times \frac{W}{2}\}$	88.62%	95.52%	99.53%	94.56%
		$\{\frac{H}{4} \times \frac{W}{4}\}$	81.79%	87.27 %	97.51%	88.86%

Table 3. Performance with different types of bypass. "–" denotes no bypass, and the "Residual" means the residual feature bypass started at the input image, while the "Residual-1st" method has a residual bypass started at the output of the first transformer block. The experiments are conducted with a Transformer-based model with three blocks, and the resolution of features after the first block is $\{H, W\}$.

Bypass	IoU(PT)	IoU(HT)	IoU(BG)	mIoU
–	**92.59%**	98.0%	99.83%	96.81%
Residual	92.57%	98.02%	99.85%	96.81%
Residual-1st	92.7%	**98.06%**	99.85%	**96.87%**
FFP	90.7%	96.8%	99.78%	95.76%

Effect of the Loss Function. We compared Cross Entropy Loss and our loss with the baseline model. Results in Table 4 show that our loss function combining weighted dice loss and weighted focal loss has improved performance.

Table 4. We compared the different loss functions with the configuration of 3 blocks and $\{H \times W\}$ resolution.

Loss Function	IoU(PT)	IoU(HT)	IoU(BG)	mIoU
Cross Entropy Loss	92.59%	98.0%	99.83%	96.81%
Focal Loss	91.86%	97.78%	99.82%	96.47%
Weighted Focal Loss	92.43%	97.92%	99.84%	96.73%
Weighted Dice Loss+Weighted Focal Loss(Ours)	**92.65%**	**98.03%**	**99.86%**	**96.85%**

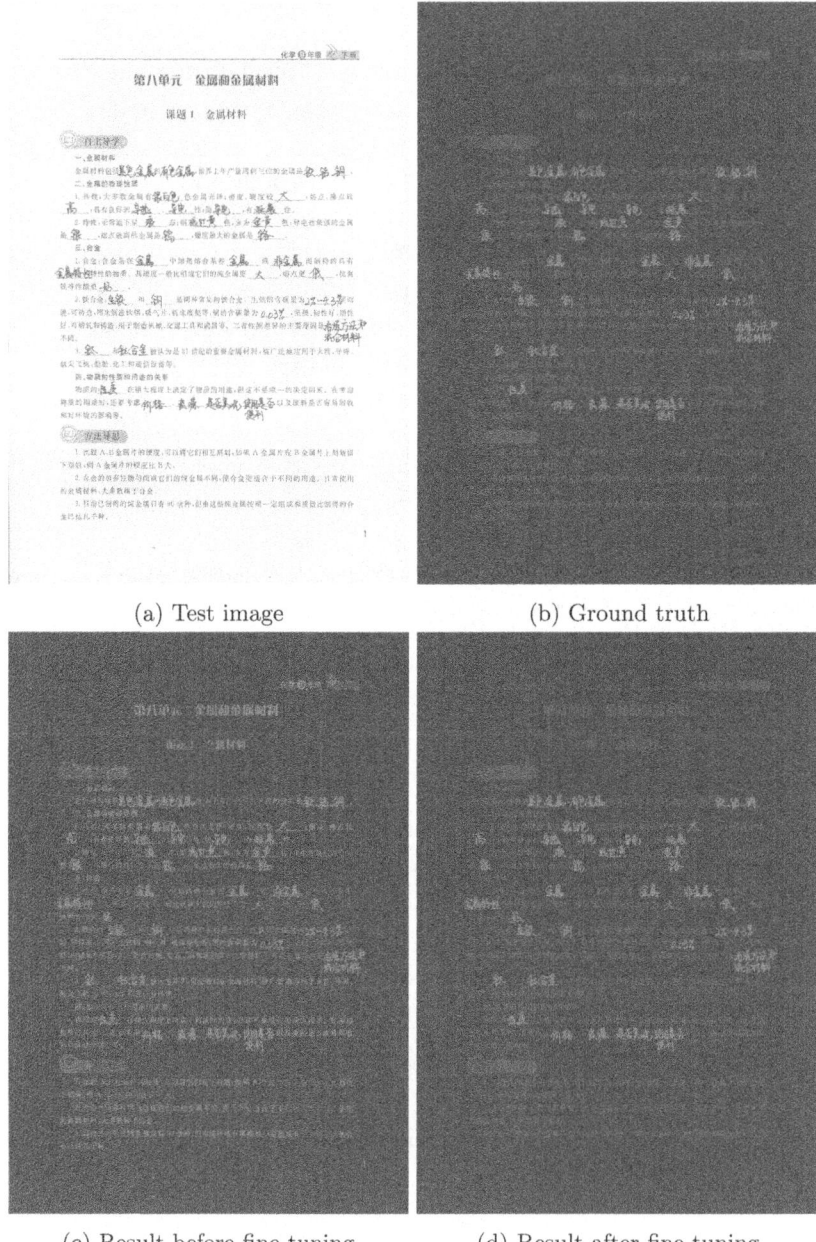

Fig. 6. The test result before and after fine-tuning the model on the training set of the Chinese dataset.

4.3 Experimental Results on the Self-Built Chinese Document Image Dataset

We have built a Chinese dataset with pixel-level annotation, including samples from student exercise books and examination papers. Each original document image has a size of $\{2467 \times 3472\}$ and is divided into samples with image size of $\{256 \times 256\}$. There are 840 samples in total, with 700 samples for training, and 140 samples for testing. We trained the model for the other 4000 iterations, and the design of other hyper-parameters is the same as the pertaining network. After fine-tuning, the IoU score is $\{99.66\%, 91.64\%, 90.33\%\}$ for BG, PT, and HT, and the mIoU is 93.88% on the test sample. The visualization of test results is shown in the Fig. 6.

5 Conclusion

This paper presents a deep-learning-based image semantic segmentation network for handwritten and printed text separation in document images at the pixel level. The network structure includes a transformer-based encoder, a light-weight CNN-based decoder, and a residual feature bypass. A loss function is designed by fusing both focal loss and dice loss. Experimental conducted on the public SignaTR6K dataset and a self-built dataset show that the proposed method is effective. Future work will explore image segmentation methods to better use the spatial context constraints in complex document images.

References

1. Vafaie, M., Bruns, O., Pilz, N., Waitelonis, J., Sack, H.: Handwritten and printed text identification in historical archival documents. In: Archiving Conference, vol. 19, pp. 15–20. Society for Imaging Science and Technology (2022)
2. Dhar, D., Garain, A., Singh, P.K., Sarkar, R.: Hp_docpres: a method for classifying printed and handwritten texts in doctor's prescription. Multimedia Tools Appl. **80**, 9779–9812 (2021). https://doi.org/10.1007/s11042-020-10151-w
3. Norkute, M., Herger, N., Michalak, L., Mulder, A., Gao, S.: Towards explainable AI: assessing the usefulness and impact of added explainability features in legal document summarization. In: Extended Abstracts of the 2021 CHI Conference on Human Factors in Computing Systems, pp. 1–7 (2021)
4. Subramani, N., Matton, A., Greaves, M., Lam, A.: A survey of deep learning approaches for ocr and document understanding. arXiv preprint arXiv:2011.13534 (2020)
5. Huang, L., et al.: EnsExam: a dataset for handwritten text erasure on examination papers. In: Fink, G.A., Jain, R., Kise, K., Zanibbi, R. (eds.) Document Analysis and Recognition - ICDAR 2023, ICDAR 2023, LNCS, vol. 14189, pp. 470–485. Springer, Cham (2023). https://doi.org/10.1007/978-3-031-41682-8_29
6. Badrinarayanan, V., Kendall, A., Cipolla, R.: Segnet: a deep convolutional encoder-decoder architecture for image segmentation. IEEE Trans. Pattern Anal. Mach. Intell. **39**(12), 2481–2495 (2017)

7. Dutly, N., Slimane, F., Ingold, R.: PHTI-WS: a printed and handwritten text identification web service based on FCN and CRF post-processing. In: 2019 International Conference on Document Analysis and Recognition Workshops (ICDARW), vol. 2, pp. 20–25 (2019)
8. Gholamian, S., Vahdat, A.: Handwritten and printed text segmentation: a signature case study. In: 2023 IEEE/CVF International Conference on Computer Vision (ICCV) (2023)
9. Xie, E., Wang, W., Yu, Z., Anandkumar, A., Alvarez, J.M., Luo, P.: Segformer: simple and efficient design for semantic segmentation with transformers. NIPS **34**, 12077–12090 (2021)
10. Zagoris, K., Pratikakis, I., Antonacopoulos, A., Gatos, B., Papamarkos, N.: Handwritten and machine printed text separation in document images using the bag of visual words paradigm. In: ICFHR, pp. 103–108 (2012)
11. Belaïd, A., Santosh, K., Poulain d'Andecy, V.: Handwritten and printed text separation in real document. In: The Thirteenth IAPR International Conference on Machine Vision Applications (2013)
12. Dutly, N., Slimane, F., Ingold, R.: PHTI-WS: a printed and handwritten text identification web service based on FCN and CRF post-processing. In: 2019 International Conference on Document Analysis and Recognition Workshops (ICDARW), vol. 2, pp. 20–25 (2019). https://doi.org/10.1109/ICDARW.2019.10033
13. Jo, J., Koo, H.I., Soh, J.W., Cho, N.I.: Handwritten text segmentation via end-to-end learning of convolutional neural networks. Multimedia Tools Appl. **79**, 32137–32150 (2020)
14. Prikhodina, A.: Handwritten and Printed Text Separation in Historical Documents. Ph.D. thesis, Karlsruher Institut für Technologie (KIT) (2022)
15. Dosovitskiy, A., et al.: An image is worth 16 × 16 words: Transformers for image recognition at scale. CoRR **abs/2010.11929** (2020). https://arxiv.org/abs/2010.11929
16. Lin, T.Y., Goyal, P., Girshick, R., He, K., Dollár, P.: Focal loss for dense object detection. IEEE Trans. Pattern Anal. Mach. Intell. **PP**(99), 2999–3007 (2017)
17. Sudre, C.H., Li, W., Vercauteren, T., Ourselin, S., Jorge Cardoso, M.: Generalised dice overlap as a deep learning loss function for highly unbalanced segmentations. In: Cardoso, M.J., et al. (eds.) DLMIA/ML-CDS -2017. LNCS, vol. 10553, pp. 240–248. Springer, Cham (2017). https://doi.org/10.1007/978-3-319-67558-9_28
18. Contributors, M.: MMSegmentation: Openmmlab semantic segmentation toolbox and benchmark. https://github.com/open-mmlab/mmsegmentation (2020)
19. Long, J., Shelhamer, E., Darrell, T.: Fully convolutional networks for semantic segmentation. In: CVPR, pp. 3431–3440 (2015)

Ablation Study of a Multimodal Gat Network on Perfect Synthetic and Real-world Data to Investigate the Influence of Language Models in Invoice Recognition

Lukas-Walter Thiée

Leuphana University Lüneburg, Lüneburg, Germany
lukasthiee@web.de

Abstract. Document analysis and invoice recognition have been significantly advanced in recent years by grid-based, graph-based and transformer architectures. However, it is not only the model architecture that influences an approach's results, but also the quality of training and test data. In this paper, we perform an ablation study on an existing state-of-the-art pre-trained multimodal GAT network. Therein we investigate two kinds of modifications to understand the sensitivity of the results by (1) exchanging the language module and (2) applying both the original and modified network on a perfect synthetic and an imperfect real-world dataset. The results of the study show the importance of language modules for semantic embeddings in multimodal invoice recognition and illustrate the impact of data annotation quality. We further contribute an adapted GAT model for German invoices.

Keywords: Invoice recognition · GAT · Synthetic data · Inv3D · GraphDoc

1 Introduction

In the digital age, businesses are increasingly reliant on data-driven processes to streamline operations and enhance efficiency. Among these, automated invoice processing stands out as a critical component, playing a pivotal role in the financial and administrative functions of organizations. The ability to swiftly and accurately handle invoices not only accelerates payment cycles but also reduces errors associated with manual data entry, thereby optimizing resource allocation and improving overall financial management. Using structured invoice data, various stakeholders can obtain information and optimize processes, e.g., for financial auditing [1].

Despite the potential for electronic transmission and standardized integration of invoice data [4], the prevailing practice, in both B2B and B2C relationships, remains sending invoices in paper or pdf format, necessitating the extraction of information from the document or file once again. Leveraging structured, meta, or blockchain data is not common practice.

Characteristics of Invoice Recognition. Automated invoice processing involves the extraction and interpretation of pertinent information from a myriad of invoice documents. This task, while crucial, is often challenging due to the inherent variability in invoice formats, diverse language structures, and the complexity of interrelations between different data elements. In contrast to pure, sequential text, invoices represent a form of visually rich documents [5], on which various signals, especially semantics and 2-dimensional layout, contain the information. The plethora of information on invoices can range from key items like addresses, dates, and invoice numbers, over account and payment details, to product quantities and descriptions (line items). Prior research has investigated the information types and pipeline challenges in information extraction from invoices [6].

Due to the privacy of sensitive personal and financial data, another hurdle in invoice recognition research is to get access to datasets that are sufficient in annotation quality and sample quantity to train machine learning algorithms. Existing publicly available datasets, such as SROIE[1] or FUNSD,[2] are very beneficial for the research community and used for various benchmark comparisons. However, these are usually tailored to a specific domain, contain too few samples, only consider a certain set of class labels, or generally have poor annotation quality. Therefore, it is promising to utilize a combination of synthetic and real data to improve both annotation quality and sample size, such as the *Document Information Localization and Extraction* (DocILE[3]) dataset.

Conventional methods, such as rule-based systems and traditional machine learning models, have demonstrated limitations in handling the intricate nature of invoice data, prompting the exploration of advanced techniques capable of capturing nuanced relationships within these documents. In order to effectively capture information and address the challenges posed by the abundance of unstructured data and diverse layouts, a multitude of new approaches within the field of machine learning (ML) have been introduced in recent years. These approaches encompass various techniques, including NLP-based models [7, 8], computer vision methodologies [9], graph-based methods [5, 10, 11], deep learning networks [12, 13], and transformer architectures [14–16]. One such promising avenue is the application of Graph Attention Networks (GATs), a subset of neural networks specifically designed to model complex relationships within graph-structured data. Graphs, in this context, represent the intricate network of connections between various entities present in an invoice, such as vendor details, line items, and transaction amounts. The inherent ability of GATs to weigh and prioritize different elements of the graph based on contextual relevance makes them a compelling choice for automated invoice processing [17].

Research Questions and Goal. Current research points to the benefits of four application streams, 1. Fine-tuning of pre-trained models, 2. Designing multimodal inputs, 3. Integrating (large) language models, and 4. Utilizing synthetic data. This research paper wants to leverage all of these streams and delves into the application of Graph Attention Networks to two distinct datasets, each presenting unique challenges and characteristics in the realm of automated invoice processing. By harnessing the power of GATs, we

---
[1] SROIE: https://rrc.cvc.uab.es/?ch=13
[2] FUNSD: https://guillaumejaume.github.io/FUNSD
[3] DocILE: https://docile.rossum.ai/

aim to enhance the accuracy and efficiency of information extraction from invoices, ultimately contributing to the broader goal of seamless and error-free automated financial workflows.

In particular, we perform a modified ablation study on an existing GAT approach. Whereas a conventional ablation study would omit parts of the model to understand the contribution of the component, we exchange an English word model with a German word model. We then apply both models on two different datasets to understand the impact of input data quality. To this end, we leverage two state-of-the-art approaches for our study both related to the *17th International Conference on Document Analysis and Recognition* (ICDAR). Inv3D by Hertlein et al. [18] was presented at ICDAR 2023[4] and GraphDoc by Zhang et al. [19] run the leaderboard in the ICDAR 2023 Competition on DocILE.[5] We strongly believe it is imperative that not only the models of state-of-the-art approaches, but also the entire ML pipeline, including the original and additional datasets, are validated to ultimately minimize ML-related debt, such as reproducibility debt and data quality debt, as introduced by Sculley et al. [20]. The approaches selected for this study have been chosen to represent state-of-the-art in invoice recognition and dataset synthesis. They will be briefly described in our related work section. The comparative analysis of GATs on the datasets will provide insights into the adaptability and generalizability of the proposed approach across different domains. As we embark on this research journey, the following key questions will guide our investigation:

1. What performance impact does the match between a model's language module and the dataset's language have?
2. How does the performance of the model change when switching from perfect (synthetic) to imperfect (real-world) data?

Relevance and Contribution. Addressing these questions will not only contribute to the advancement of automated invoice processing but will also shed light on the broader applicability and limitations of Graph Attention Networks in handling complex and dynamic relational data structures. Through this research, we aim to provide valuable insights that will inform the development of more robust and adaptive solutions for automated invoice processing in diverse business environments. We contribute a performance test of a multimodal state-of-the-art approach, to analyze the impact of language models. Furthermore, we extend GraphDoc with an existing German language transformer model.

2 Related Work: Models and Datasets

2.1 GraphDoc

Zhang et al. present *"a multimodal graph attention-based model for various document understanding tasks"* [19], called GraphDoc.[6] GraphDoc leverages three multimodal inputs to the network, namely positional, textual and image information, and unites

---
[4] https://icdar2023.org/program/accepted-papers/
[5] https://icdar2023.org/program/competitions/
[6] https://github.com/ZZR8066/GraphDoc

those in a gate fusion layer. The textual features are integrated via the pre-trained English Sentence-BERT (ESB) model as a language module, which provides sentence embeddings. The weights of the language module are not being pre-trained in GraphDoc. Zhang et al. build upon the idea that individual elements of a document can depend on their direct neighborhood, and therefore *"inject the graph structure into the attention mechanism to form a graph attention layer so that each input node can only attend to its neighborhoods"* (see Fig. 1). GraphDoc is pre-trained on a sample of 320k images from the RVL-CDIP dataset to learn document representations through masked sentence modelling, and it is evaluated on the FUNSD dataset. GraphDoc is based on regions of interest (RoI), which represent contiguous areas of the layout of a document. EasyOCR is used to obtain texts, bounding boxes, and regions. The structure of the document is induced by the graph attention mechanism so that k = 36 of the nearest neighbors of a region are taken into account. Zhang et al. not only report very good results of the approach on various benchmark datasets in their paper, but also lead the ranking in the DocILE challenge by Šimsa et al. [21]. Table 1 lists selected results.

Table 1. Collection of F1-scores of GraphDoc on different datasets [19].

Dataset Model	FUNSD (Form) F1	SROIE (Receipt) F1	CORD (Receipt) F1	DocILE (Invoice) F1
GraphDoc (ESB)	0.8777	0.9845	0.9693	0.7425[7]

2.2 (German) BERT

With the rise of vectorized word models in natural language processing, like Word2Vec [22] or GloVe [23], also transformer architectures were introduced to represent language. One of those architectures is BERT (Bidirectional Encoder Representations from Transformers [24]), which has been developed and open-sourced by Google.[8] It *"is basically a multi-layer bidirectional transformer"* [14], to generate semantic embeddings of words within the context of its sentence. It uses the self-attention mechanism to weigh the significance of different words in a sentence. By training on large amounts of textual data, BERT learns to generate contextually rich word embeddings, allowing it to grasp the relationships between words and their contextual meanings.

English Sentence BERT is an extension of the BERT model, specifically tailored for sentence-level embeddings. It focuses on learning sentence representations. This makes it suitable for various NLP tasks such as sentence similarity, classification, and semantic textual similarity. On the other hand, the German BERT model[9] by deepset.ai [25] is a BERT-based model fine-tuned for the German language. Fine-tuning involves

---

[7] Robust Reading Competition: DocILE 2023, Key Information Localization and Extraction: https://rrc.cvc.uab.es/?ch=26&com=evaluation&task=1
[8] https://ai.googleblog.com/2018/11/open-sourcing-bert-state-of-art-per.html
[9] https://www.deepset.ai/german-bert

training a pre-trained model on a specific task or dataset to adapt it for domain-specific or language-specific applications. The German BERT model captures the intricacies of the German language and can be employed in a range of NLP tasks for German text.

2.3 Inv3D, Perfect Data

Hertlein et al. [18] establish a new benchmark dataset called Inv3D[10] for automated invoice processing, mainly targeted at dewarping invoice images and invoice recognition tasks. Their large-scale high-resolution dataset consists of 25,000 samples including the flatbed image and pdf file and two ground-truth annotation json-files, i.e., complete list of words and relevant areas (key and line items), split into 0.7 train, 0.15 validation, and 0.15 test sets. *"The dataset creation pipeline consists of four stages: resource preparation, invoice rendering, invoice warping, and finally the auxiliary map generation"* [18]. For the purpose of our study the first two stages are relevant. Inv3D is based on 100 real template layouts. They leverage the python Faker package to randomly generate coherent invoice content, e.g., sales orders and personas, and apply random modifications to the design of the invoice, e.g., font sizes and colors, and exchange company logos. The content is mapped to a machine-readable tag structure, such as *buyer.shipment.address*. The train dataset contains 162 classes. They also provide the code of their pipeline to generate further samples with new content, layout, and other types of documents. We utilize Inv3D in this study, because the ground-truth of this dataset is perfect, meaning the annotations and boxes are unambiguous, which is crucial for understanding class based inference on token level.

2.4 Private Dataset (PD), Imperfect Data

We have a non-public data set available that consists of German invoices and receipts, hereinafter referred to as PD. The data originates from the accounting systems of different German agricultural businesses. These invoices encompass a diverse range of transactions and commodities, extending beyond exclusively agricultural products. Instead, they comprise a substantial number of service-related invoices and invoices associated with the acquisition of diverse industrial and consumer goods. The pdf files correspond to scanned invoices sourced from different companies and various scanning devices, leading to variations in scan quality. In total, our dataset comprises 977 invoices originating from 494 distinct vendors and involving 531 distinct recipients. Abbyy Finereader OCR results in 196,548 (43,621 distinct) tokens, with a mean average of 201 tokens per invoice. The results are subject to various challenges, such as the umlauts in the German language or unstandardized abbreviations. The invoices have been annotated for 67 different classes, such as *adresse, beleg_datum,* and *rechnung_nr*, but also *steuer_nr, iban,* and *ust_id_number*, which are more specific to the German-speaking area (see Table 3). The annotation is based on a sophisticated rule-based algorithm, which yields decent, however, not perfect results. Annotations could be inaccurate and incomplete. For example, they do not distinguish between vendor and recipient address, nor do they consider line items. For various tokens, the assignment between ground-truth data and

---
[10] https://felixhertlein.github.io/inv3d

OCR results is not unambiguous. Some of the documents may contain handwritten notes and they vary considerably in type and layout.

3 Experiments

Our study design consists of four experiments – two models and two datasets – each training and testing the corresponding model on one of the aforementioned datasets (see Table 2). The first model is the pre-trained model from [19], presented in Sect. 2.1 with English Sentence-BERT (EB). We use the pre-trained main network and *GraphDoc-ForTokenClassification* adapted to our experiments. This network tail is used to obtain token level classifications. As a second model we use GraphDoc again, but replace the language module with an implementation of German BERT (GB). We then fine-tune the pre-trained models on our datasets. Rather than omitting a component of the model, we exchange the component. We do this because we consider it necessary to keep the semantic module, as the model is designed for multimodal inputs. Our ablation study is therefore not designed as a hyperparameter optimization, but rather as a general model verification. We are committed to using existing models, as well as testing and expanding their performance as part of scientific rigor in business informatics. Experiment E1 and E4 are our focus experiments, as the language of the dataset and the language module within the model match, i.e., English for E1 and German for E4. The proposed experiment setup provides us with six expedient comparison pairs, two that test the influence of the language module (E1 vs. E2 and E3 vs. E4) and four that test the combined influence of the dataset quality and language fit (E1 vs. E3, E2 vs. E4, E2 vs. E3 and E1 vs. E4).

Table 2. Experimental setup and comparison options.

Model	Dataset	Inv3D (English)	PD (German)
	Samples	Train: 17.500 Test: 3.750 Classes: 7/162	Train: 782 Test: 195 Classes: 7/67
GraphDoc (English Sentence BERT)		E1	E3
GraphDoc (German BERT)		E2	E4

Expected results (a priori qualitative hypotheses). With regard to our research questions and the possibilities for comparison between the experiments, we put forward the following qualitative hypotheses for the model performance P, measured as macro average F1-score. The comparisons are illustrated with arrows in Table 2. The direction of the arrow indicates an expected relative increase in performance.

- **H1**: Isolated impact of the language module: The performance of the experiments with a matching language of the language module and dataset language is expected to be higher than the performance of experiments with a non-matching language, when keeping the dataset. This means that in the case of Inv3D we expect a higher performance with GraphDoc EB, $P_{E1} > P_{E2}$. And in the case of PD, better performance with GraphDoc GB, $P_{E3} < P_{E4}$.

- **H2**: Combined effects: Since we are not modifying the datasets themselves, we can only investigate the combined impact of data quality and language fit. Generally, we expect the performance to decrease when switching from perfect synthetic data to imperfect real-world data, $P_{E1} > P_{E3}$, $P_{E2} > P_{E3}$, and $P_{E1} > P_{E4}$. However, For E2 vs. E4 we anticipate the effect of language fit to outweigh disadvantages of data quality, hence $P_{E2} < P_{E4}$.

3.1 Preprocessing and Training

In order to train and test the models we have to prepare the datasets according to the model input structure. Inv3D contains 17.500 train and 3.750 test samples, and PD contains 782 train and 195 test samples. For both datasets we use the provided bounding boxes of the ground truth of all words and set these to the class *none*. We then match all words and bounding boxes with the class label ground truth to obtain word-level inputs.

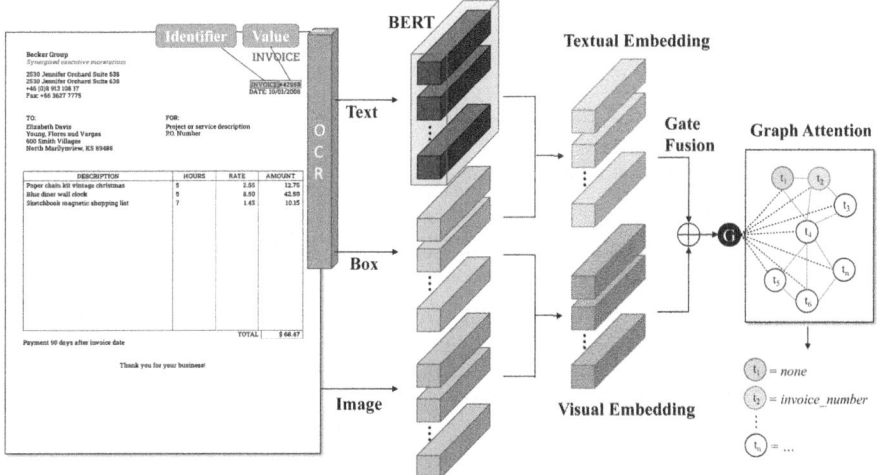

Fig. 1. Exemplary document from Inv3D [18] testset and model structure adapted from [19].

To ensure that we have the same semantic classes in both datasets, we focus on 7 selected classes, such as *invoice_date = beleg_datum, invoice_number = rechnung_nr,* and *summary.balance = betrag*. Table 3 gives an impression of the possible and used classes. Whereas PD does not distinguish seller and buyer addresses, Inv3D uses a more detailed structure of the classes (*buyer, seller,* and *beneficiary*) and also explicitly enumerates product details. With this structure Inv3D can potentially be used for line item recognition training. We apply the model hyperparameters suggested by Zhang et al. [19], batch size 4, learning rate $5 * 10^{-5}$, and fine-tune the models for 50 epochs each. Figure 1 illustrates the general process of the pipeline, i.e., word-level tokens, their different embeddings, and graph structure.

Table 3. Selection of class labels in the datasets.

Type	Inv3D classes		PD classes
Key items: contact	buyer.bill.address buyer.bill.address.city buyer.bill.address.city.postcode buyer.bill.address.street buyer.bill.contact buyer.bill.email **buyer.bill.customer_id** ② buyer.bill.phone_number buyer.company.name buyer.shipment.address buyer.shipment.contact buyer.shipment.phone_number	**seller.address** ① seller.address.city seller.address.city.postcode seller.address.street seller.contact **seller.email** ③ **seller.phone_number** ④ seller.company.name seller.company.slogan seller.fax_number seller.salesperson	**adresse** ① **telefon_nr** ④ telefax_nr **email** ③ **kunden_nr** ② bestell_nr referenz_nr hndlreg_nr steuer_nr ust_id_number url bic
	beneficiary.iban.account_code beneficiary.name beneficiary.bic beneficiary.iban	beneficiary.bank.address.city beneficiary.bank.address.street beneficiary.bank.name beneficiary.bank.address	blz konto_nr iban karten_nr
Key items: general	**summary.balance** ⑦ summary.discount summary.subtotal summary.tax_rate summary.tax_total summary.discount summary.shipping.method summary.shipping.price	**invoice_date** ⑤ invoice_due_date **invoice_number** ⑥ payment_terms	**beleg_datum** ⑤ **rechnung_nr** ⑥ **betrag** ⑦ zahlungsart referenz_nr bezahlt_flag zu_zahlen_flag globale_lokations_nr
Line items	products.0.description products.0.id products.0.tax_rate	products.0.quantity products.0.total products.0.unit_price	

* Ⓝ indicates counterparts in the datasets and integration in our train/test set

4 Results and Discussion

The overall and class-level performance metrics of the experiments are summarized in Table 4 and Table 5. We use the macro average F1-score for the evaluation of the model performance P. F1-score is the harmonic mean between precision and recall, which is evaluated per class. A macro average in multi-class classification is the arithmetic mean of all per-class scores regardless of their support, whereas a weighted average considers the class distribution.

Since we use new combinations of models and datasets, there is no direct benchmark comparison. Nevertheless, we can put the macro F1-score of the experiments in relation to results of the original model on other datasets (see Sect. 2.1 in Related Work). The F1-scores of the focus experiments, i.e., the ones with a language fit (E1 = 0.9324 and E4 = 0.8593), represent a significant improvement compared to the application of GraphDoc on the DocILE dataset (0.7425). However, in terms of dataset and class complexity,

Table 4. Classification report on Inv3D test dataset.

Model/Class Metric	Precision	Recall	F1-score
GraphDoc (English BERT): Experiment 1			
⓪ none	0.9948	0.9923	0.9935
① buyer.bill.address	0.8476	0.9434	0.8929
② buyer.bill.customer_id	0.8338	0.8576	0.8455
③ seller.email	0.9742	0.9029	0.9372
④ seller.phone_number	0.9903	0.9507	0.9701
⑤ invoice_date	0.9940	0.9327	0.9624
⑥ invoice_number	0.9538	0.9526	0.9532
⑦ summary.balance	0.9054	0.9030	0.9042
accuracy 0.9865			
macro avg.	0.9367	0.9294	**0.9324**
weighted avg.	0.9871	0.9865	0.9867
GraphDoc (German BERT): Experiment 2			
⓪ none	0.9878	0.9376	0.9621
① buyer.bill.address	0.3510	0.7992	0.4877
② buyer.bill.customer_id	0.3452	0.7898	0.4804
③ seller.email	0.9350	0.8881	0.9109
④ seller.phone_number	0.9578	0.9207	0.9389
⑤ invoice_date	0.9826	0.8806	0.9288
⑥ invoice_number	0.9395	0.9421	0.9408
⑦ summary.balance	0.8500	0.8522	0.8511
accuracy 0.9296			
macro avg.	0.7936	0.8763	**0.8126**
weighted avg.	0.9603	0.9296	0.9410

the DocILE benchmark is more complex. The performance of GraphDoc in the SROIE challenge is not achieved (0.9845). These comparisons provide a basic ranking but are limited in their informative value due to different dataset properties, such as document type, sample size, type and distribution of class labels, as well as variance of layouts. For example, more classes are considered in the DocILE challenge and fewer classes in the SROIE challenge.

All experiments show high accuracy scores. However, the classes are strongly imbalanced and the majority class (*none*) consumes the result. Nonetheless, precision, recall and F1 also achieve good scores. The difference between macro average precision and recall is balanced for all experiments, which shows that the models learn to discriminate the classes despite the class imbalance. Table 6 summarizes the macro average F1-scores of the experiments and shows the difference between the experiments in percentage points (pp). The comparison of the F1 scores serves to test our qualitative hypotheses.

Table 5. Classification report on PD test dataset.

Model/Class	Metric	Precision	Recall	F1-score
GraphDoc (English BERT): Experiment 3				
⓪ none		0.9509	0.9115	0.9308
① adresse		0.5837	0.7291	0.6484
② kunden_nr		0.5661	0.8045	0.6646
③ email		0.8516	0.8201	0.8356
④ telefon_nr		0.9133	0.8103	0.8587
⑤ beleg_datum		0.7781	0.8704	0.8217
⑥ rechnung_nr		0.7863	0.7244	0.7541
⑦ betrag		0.7388	0.8049	0.7704
accuracy	0.8838			
macro avg.		0.7711	0.8094	**0.7855**
weighted avg.		0.8967	0.8838	0.8887
GraphDoc (German BERT): Experiment 4				
⓪ none		0.9829	0.9740	0.9784
① adresse		0.8690	0.9169	0.8923
② kunden_nr		0.7914	0.8271	0.8088
③ email		0.9364	0.8571	0.8950
④ telefon_nr		0.9213	0.8410	0.8794
⑤ beleg_datum		0.8438	0.9000	0.8710
⑥ rechnung_nr		0.8099	0.7717	0.7903
⑦ betrag		0.7063	0.8211	0.7594
accuracy	0.9618			
macro avg.		0.8576	0.8636	**0.8593**
weighted avg.		0.9629	0.9618	0.9622

Table 6. Macro average F1-scores and deltas between the experiments.

Model	Dataset	Inv3D		PD	
GraphDoc (EB)		**E1**	0.9324	**E3**	0.7855
		Δ 11,98pp		Δ 7,38pp	
GraphDoc (GB)		**E2**	0.8126	**E4**	0.8593

Inv3D ↔ PD deltas: Δ 14,6pp (E1–E3), Δ 2,71pp, Δ 7,30pp, Δ 4,67pp (E2–E4).

- **H1:** $P_{E1} > P_{E2}$ and $P_{E3} < P_{E4}$. The hypothesis regarding the influence of the network's language module on performance is confirmed in both comparisons. In both cases, there is a significant delta between the experiment with a language fit and the experiment without language fit. In the case of Inv3D almost 12pp. This means that

a language module that matches the dataset has a strong positive influence on the classification results, if the dataset is maintained. It is interesting to note that the effect of the language module is stronger for synthetic data than for real-world data. Hypothesis 1 reflects the real-world use case, as the dataset is not usually exchanged, but the model is adapted.
- **H2**: $P_{E1} > P_{E3}$, $P_{E2} > P_{E3}$, and $P_{E1} > P_{E4}$. Our hypothesis about the combined impact of data quality and language fit is also confirmed. In general, better results are achieved with synthetic data. The comparison between E1 and E3 shows the largest effect, of approximately 15pp. E2 vs. E3 and E1 vs. E4 also confirm this. In the case of E2 vs. E4, where we expected the effects of language and data quality to counteract, we see that a matching language can outweigh issues with data quality, in fact $P_{E2} < P_{E4}$. However, the total delta between E2 and E4 of 4.67pp is only a third of E1 vs. E3.

As expected, the case of language fit and synthetic data achieves the best results (E1). Unfavorable influences from inferior data quality, e.g., from annotation or OCR errors, can be compensated by adjusting the network's language module. Purely synthetic data provide best training results, as these data inherit synthesized patterns. In real-world scenarios, augmenting the real-word data with synthetic data can benefit the training. As E2 and E3 also achieve good results, it shows that features from the visual embedding also have predictive value in the model, which strengthens the approach of multimodal inputs.

The confusion matrices of our focus experiments E1 (Table 7) and E4 (Table 8) show that, in general, most misclassifications occur between the majority class '*none*' and each foreground class. For example, the address tokens can be recognized well as an overall construct. This is probably due to the exposed position and size of the tokens in that class. Classes with a syntactic structure such as email, phone numbers, and dates can also be assigned well. Identifier classes, such as invoice numbers or customer ID, also show good classification results in our experiments. This is due to the attention mechanism in the network. The weights in the GAT layers for these tokens show a strong influence of their neighborhood context [10]. This means that the actual invoice number is influenced by the identifier, also represented in Fig. 1. The total amount class (*summary.balance* and *betrag*) show confusion with the *none* class. This could be due to the fact that several amounts can be found on invoices, but these were labeled as '*none*'.

Discussion and Future Work. Our ablation study shows that language modules within a network, that featurizes semantic embeddings, have a strong impact on performance. Overall, the multimodal approach, which combines semantic embeddings, image elements, and layout structure, shows promising results on both datasets. The approach of using the pre-trained models also proves to be expedient, as good results are consistently achieved. Nevertheless, our training is a fine-tuning, so that a full training including the optimization of hyperparameters of the models can possibly lead to even better results in the future.

Although the original GraphDoc model is designed for RoI, i.e., paragraphs, text blocks, tables, etc., it also has good performance in the case where "*both the region-level and word-level boxes are the same*" [19]. This means that modeling at token level is perfectly permissible. However, it is to be expected that preprocessing RoI instead of

Table 7. Confusion matrix of E1.

Actual		Prediction								Support
		⓪	①	②	③	④	⑤	⑥	⑦	
none	⓪	374502	2401	108	12	25	11	10	345	377414
seller.address	①	585	13899	22	8	2	4	2	211	14733
b.b.customer_id	②	71	34	873	2	9	5	3	21	1018
seller.email	③	99	4	1	1097	0	0	4	10	1215
s.phone_number	④	244	3	5	2	5923	0	45	8	6230
invoice_date	⑤	306	10	8	3	3	5969	0	101	6400
invoice_number	⑥	97	10	18	2	17	6	3094	4	3248
summary.balance	⑦	572	38	12	0	2	10	86	6700	7420

Table 8. Confusion matrix of E4.

Actual		Prediction								Support
		⓪	①	②	③	④	⑤	⑥	⑦	
none	⓪	31574	711	7	4	12	41	7	62	32418
adresse	①	405	4765	6	6	1	2	3	9	5197
kunden_nr	②	12	2	110	0	2	0	5	2	133
email	③	20	5	0	162	0	0	0	2	189
telefon_nr	④	49	0	8	0	328	0	4	1	390
beleg_datum	⑤	19	0	4	1	0	243	0	3	270
rechnung_nr	⑥	16	0	4	0	4	0	98	5	127
betrag	⑦	29	0	0	0	9	2	4	202	246

word-level tokens can achieve even better results. This would also support the actual approach of the BERT models, as they should embed the sentence context and not only single words. Our approach of measuring the influence of data quality via the exchange of the dataset is only valid to a limited extent. With a different approach, e.g., by manipulating the labels in the original data set, this effect could be investigated in more isolated form. Nevertheless, the observation that in one of our experimental comparisons the effect of language fit outweighs inferior data quality (E2 vs. E4) is quite interesting. This effect indicates that the capabilities for generating automatic embeddings can partly compensate issues with data quality. Nevertheless, annotation and preprocessing quality must still be assigned a high level of importance.

There are multiple possibilities to continue this study. One option is to integrate further class labels of the datasets, e.g., for line-item recognition and to generate further synthetic documents via the Inv3D pipeline, so that the variance of the dataset increases.

An investigation of other language modules, such as GBERT and GELECTRA [26], could provide further insights into the influence of semantic embeddings in invoice recognition and also promise even better performance, as they achieve state-of-the-art performance across document classification and named entity recognition tasks. Also, manipulating the datasets (annotation, text, boxes), and investigating different entity levels (character, word, region) with different OCR engines could yield further insights.

While the confirmation of the hypotheses appears to be straight-forward, it is nevertheless important to deal with the conclusions drawn from them. It means, for example, that competitions with a generally higher data quality show better results. However, since models should be valid beyond the benchmark dataset, a validation on unseen data from different datasets is essential – considering consistent requirements for comparison, e.g., number and type of prediction classes. The key takeaway from our experiments is that the right choice of language model can partially compensate for problems with data quality. The multimodal combination of specialized models, e.g., (large) language models for semantic tasks, graphs for structure related tasks, or convolutional networks for image tasks, is a promising research path. We therefore argue that the use of multimodal and multi-model approaches, in the sense of ensemble learning, can achieve highest generalizability on unseen datasets.

5 Conclusion

In this paper, we conduct an ablation study in which we exchange the language module (BERT) of a state-of-the-art document understanding model, namely multimodal pretrained GraphDoc. The original and the modified model are fine-tuned on two different datasets, a perfect synthetic (Inv3D) and a real-world dataset (PD). The classification metrics achieve good results comparable to other benchmark approaches. Our hypotheses that both a matching language model and the annotation quality have a significant influence on macro F1-score performance are confirmed. Our contribution is twofold. Firstly, we analyze and confirm the performance of a multimodal state-of-the-art approach. We show the impact of matching the language of the semantic module to the dataset language. Secondly, we extend the approach with a German language model.

Disclosure of Interests. The authors have no competing interests to declare that are relevant to the content of this article.

References

1. Krieger, F., Drews, P.: Leveraging big data and analytics for auditing: towards a taxonomy. In: ICIS 2018 Proceedings (2018)
2. Liu, Y., et al.: RoBERTa: a robustly optimized BERT pretraining approach (2019)
3. Sanh, V., Debut, L., Chaumond, J., Wolf, T.: DistilBERT, a distilled version of BERT: smaller, faster, cheaper and lighter (2019)
4. Klein, B., Agne, S., Dengel, A.: Results of a study on invoice-reading systems in Germany. In: Marinai, S., Dengel, A.R. (eds.) Document Analysis Systems VI, DAS 2004, LNCS, vol. 3163, pp. 451–462. Springer, Berlin (2004). https://doi.org/10.1007/978-3-540-28640-0_43

5. Liu, X., Gao, F., Zhang, Q., Zhao, H.: Graph convolution for multimodal information extraction from visually rich documents (2019)
6. Thiée, L.-W., Krieger, F., Funk, B.: Extraction of information from invoices – challenges in the extraction pipeline. In: Klein, M., Krupka, D., Winter, C., Wohlgemuth, V. (eds.), Informatik 2023. Designing Futures: Zukünfte gestalten ; Tagung vom 26–29 2023, Berlin. Gesellschaft für Informatik, Bonn (2023)
7. Palm, R.B., Winther, O., Laws, F.: CloudScan - a configuration-free invoice analysis system using recurrent neural networks. In: 12th International Conference on Document Analysis and Recognition (2013)
8. Singh, S.: Natural language processing for information extraction, Australia (2018)
9. Davis, B., Morse, B., Cohen, S., Price, B., Tensmeyer, C.: Deep visual template-free form parsing. In: 15th International Conference on Document Analysis and Recognition (2019)
10. Krieger, F., Drews, P., Funk, B., Wobbe, T.: Information extraction from invoices: a graph neural network approach for datasets with high layout variety. Wirtschaftsinformatik 2021 Proceedings (2021)
11. Lohani, D., Belaïd, A., Belaïd, Y.: An invoice reading system using a graph convolutional network. In: Carneiro, G., You, S. (eds.) Computer Vision - ACCV 2018 Workshops, ACCV 2018, LNCS, vol. 11367, pp. 144–158. Springer, Cham (2019). https://doi.org/10.1007/978-3-030-21074-8_12
12. Yang, X., Yumer, E., Asente, P., Kraley, M., Kifer, D., Giles, C.L.: Learning to extract semantic structure from documents using multimodal fully convolutional neural network (2017)
13. Palm, R.B., Laws, F., Winther, O.: Attend, copy, parse -- end-to-end information extraction from documents. In: ICDAR (2019)
14. Xu, Y., Li, M., Cui, L., Huang, S., Wei, F., Zhou, M.: LayoutLM: pre-training of text and layout for document image understanding (2020)
15. Garncarek, Ł, et al.: LAMBERT: layout-aware language modeling for information extraction. Doc. Anal. Recogn. ICDAR **2021**(12821), 532–547 (2021)
16. Xu, Y., et al.: LayoutLMv2: multi-modal pre-training for visually-rich document understanding (2020)
17. Veličković, P., Cucurull, G., Casanova, A., Romero, A., Liò, P., Bengio, Y.: Graph attention networks (2017)
18. Hertlein, F., Naumann, A., Philipp, P.: Inv3D: a high-resolution 3D invoice dataset for template-guided single-image document unwarping - Meta data. Karlsruhe Institute of Technology (2023)
19. Zhang, Z., Ma, J., Du Jun, Wang, L., Zhang, J.: Multimodal pre-training based on graph attention network for document understanding (2022)
20. Sculley, D., et al.: Hidden technical debt in machine learning systems. In: Advances in Neural Information Processing Systems, vol. 28 (2015)
21. Šimsa, Š., et al.: DocILE benchmark for document information localization and extraction (2023)
22. Mikolov, T., Chen, K., Corrado, G., Dean, J.: Efficient estimation of word representations in vector space (2013)
23. Pennington, J., Socher, R., Manning, C.: Glove: global vectors for word representation. In: Proceedings of the 2014 Conference on Empirical Methods in Natural Language Processing (EMNLP), pp. 1532–1543 (2014)
24. Devlin, J., Chang, M.-W., Lee, K., Toutanova, K.: BERT: pre-training of deep bidirectional transformers for language understanding (2018)
25. deepset.ai: German BERT. https://www.deepset.ai/german-bert. Accessed 07 Jan 2024
26. Chan, B., Schweter, S., Möller, T.: German's next language model (2020)

Author Index

A
Akundi, Prathyusha II-175
Allen, Jonathan Parkes II-87
Anger, Jérémy I-40
Anquetil, Eric I-3
Armenakis, Yiannis I-103
Atamni, Nour II-119
Aubry, Mathieu II-3

B
Berg-Kirkpatrick, Taylor II-87
Bertini, Marco I-154
Biescas, Nil I-27
Biondi, Niccolò I-154
Biswas, Sanket I-27
Bizais-Lillig, Marie II-37
Borkar, Jaydeep II-57
Bottaioli, Natalia I-40
Burie, Jean-Christophe I-198

C
Campaioli, Irene I-154
Camps, Jean-Baptiste II-140
Chao, Yu II-184
Chavallard, Pauline II-163
Chen, Danlu II-87
Chen, Yung-Hsin I-12

D
De Gregorio, Giuseppe II-71, II-102
Decours-Perez, Aliénor II-22
Dupin, Boris II-37

E
El-Sana, Jihad II-119

F
Facciolo, Gabriele I-40
Ferretti, Lavinia II-102
Fitsilis, Fotios I-103

G
Gardella, Marina I-40
Gatos, Basilis I-103
Georgoulea, Maria-Eleni I-103

I
Imbert, Florent I-3
Iwata, Motoi I-216

K
Kaddas, Panagiotis I-103
Kamilaki, Maria I-89
Karatzas, Dimosthenis I-154
Katsouros, Vassilis I-138
Kermorvant, Christopher I-40
Kiousi, Eleni I-103
Kise, Koichi I-216
Klut, Stefan I-73
Konstantinidou, Maria II-102
Koornstra, Tim I-73
Kouletou, Eleanna I-138
Kubade, Ashish II-175
Kyrkos, Charalambis I-103

L
Liu, Changsong II-184
Lladós, Josep I-27

M
Maas, Martijn I-73
Madi, Boraq II-119
Marthot-Santaniello, Isabelle II-71, II-102
Mikros, George I-103
Mohd, Bilal Arif Syed II-175
Morel, Jean-Michel I-40
Mouchère, Harold II-71
Mowlavi, Seginus I-40
Murel, Jacob I-125, II-87

N
Nardoni, Mariateresa I-154

O
Okamoto, Keito I-216
Oparnica, Milena I-59

P
Palaiologos, Konstantinos I-103
Papavassiliou, Vassilis I-138
Pavlopoulos, John II-102
Pena, Rodrigo C. G. II-71, II-102
Peng, Liangrui II-184
Perrin, Simon II-71
Peters, Luke I-73
Petit, Samuel I-198
Pilligua, Maria I-27
Preciozzi, Javier I-40

R
Rabaev, Irina II-119
Real, Thibaud II-163
Rigaud, Christophe I-198
Rozenberg, Olivier I-103

S
Sezgin, Tevfik Metin I-168, I-231, I-257
Shahid, Taimoor II-87
Siglidis, Ioannis II-3
Smith, David A. I-125, II-57, II-87
Soullard, Yann I-3

Soykan, Gürkan I-168, I-231, I-257
Ströbel, Phillip B. I-12
Stutzmann, Dominique II-3

T
Tadros, Antoine I-40
Tarride, Solène I-40
Tasouli, Christina I-103
Tavenard, Romain I-3
Thiée, Lukas-Walter II-199
Tsitrinovich, Vasily II-119

V
Valveny, Ernest I-27
van Koert, Rutger I-73
Vasyutinsky-Shapira, Daria II-119
Vazquez-Corral, Javier I-27
Vidal-Gorène, Chahan II-22, II-37, II-140
Vivoli, Emanuele I-154
Vlachou-Efstathiou, Malamatenia II-3
von Gioi, Rafael Grompone I-40

W
Wang, Yanwei II-184

Y
Yuret, Deniz I-168, I-231, I-257

Z
Zhang, Xiang II-87

SPRINGER NATURE

GPSR Compliance

The European Union's (EU) General Product Safety Regulation (GPSR) is a set of rules that requires consumer products to be safe and our obligations to ensure this.

If you have any concerns about our products, you can contact us on ProductSafety@springernature.com

In case Publisher is established outside the EU, the EU authorized representative is:

Springer Nature Customer Service Center GmbH
Europaplatz 3
69115 Heidelberg, Germany

The manufacturer's authorised representative in the EU is Springer Nature Customer Service Centre GmbH, Europaplatz 3, 69115 Heidelberg, Germany. If you have any concerns regarding our products, please contact ProductSafety@springernature.com

Printed and bound by CPI Group (UK) Ltd, Croydon, CR0 4YY

25/03/2026

02078187-0009